T0041462

Notting Hill Editions is an independent British publisher. The company was founded by Tom Kremer (1930–2017), champion of innovation and the man responsible for popularising the Rubik's Cube.

After a successful business career in toy invention Tom decided, at the age of eighty, to fulfil his passion for literature. In a fast-moving digital world Tom's aim was to revive the art of the essay, and to create exceptionally beautiful books that would be lingered over and cherished.

Hailed as 'the shape of things to come', the family-run press brings to print the most surprising thinkers of past and present. In an era of information-overload, these collectible pocket-size books distil ideas that linger in the mind.

nottinghilleditions.com

Andrew Jamieson trained at the Bath Centre for Psychotherapy and Counselling and received an MA in Humanistic and Integrative Psychotherapy at Middlesex University. He lectures and writes articles on a series of subjects including psychotherapy's interconnection with philosophy, music and literature. Parallel to his psychotherapeutic career, he has promoted orchestral concerts throughout the UK for over forty years.

MIDLIFE

Humanity's Secret Weapon

–

Andrew Jamieson

Notting Hill Editions

Published in 2022
by Notting Hill Editions Ltd
Mirefoot, Burneside, Kendal LA8 9AB

Series design by FLOK Design, Berlin, Germany
Cover design by Tom Etherington
Creative Advisor: Dennis PAPHITIS

Typeset by CB Editions, London
Printed and bound by Memminger MedienCentrum,
Memmingen, Germany

A CIP record for this book is available from the British Library.

ISBN 978-1-912559-38-1

nottinghilleditions.com

To Julia, Anna, Lucy and Olivia

Contents

– Introduction –

Since I began practising as a psychotherapist, three quarters of my clients have been between the age of thirty-five and fifty-five. They invariably arrive for their first session in a state of depression, anxiety and uncertainty, often unsure whether to give up a profession or a marriage. In this state of near breakdown, they feel overwhelmed by some insoluble, intractable problem. As I have traversed this challenging emotional rockface with client after client I am, time and again, impressed by how these periods of inner turmoil provide us with an unmatched opportunity to review our lives and explore our personalities. We can then attempt to adapt and reshape those aspects of our nature which constrict our development, that hold back our true potential and impede our sense of well-being.

This rite of passage that we have come to call 'the midlife crisis' has an ancient provenance. Indeed, it is the subject of the second story ever told in the earliest stirrings of Western culture. Homer's *Odyssey* recounts Odysseus's journey home to Ithaca after the end of the Trojan Wars. As the gods test his resolve, Odysseus is beset by a multitude of calamities and temptations that – during the course of his long voyage home – transform him from a youthful warrior into a wise and enlightened elder.

Several thousand years later another sacred text recounts a similar rite of passage undertaken by Dante Alighieri whose *Divine Comedy* begins with its famous opening sentence: 'Midway through this life upon which we are bound, I woke to find myself in a dark wood, where the right road was wholly lost and gone.' This epic story describes Dante's journey through the Inferno, Purgatory and finally into Paradise, in the company of his guide and mentor Virgil and then his muse Beatrice.

These confrontations with inner demons and outer misfortune experienced by Odysseus and Dante have numerous parallels in Western religious texts and literature. They appear in the Old Testament stories of Jonah and Job, in Christ's forty days and nights in the wilderness, in St John of the Cross's Dark Night of the Soul, in Goethe's drama *Faust*, in Tolstoy's accounts of Levin and Pierre's tribulations and in Eliot's *East Coker*. The sense that inner wisdom or a feeling of enlightenment and psychological repose can only be achieved through a crisis, an ordeal or a long hazardous journey is a belief that runs through the entire tradition of Western literature, mythology and spirituality.

As with Odysseus and Dante – and as with the heroes and protagonists of Goethe, Tolstoy and Eliot –

many of my clients who experience this ordeal meet the challenge and undergo some form of transformation. As they recover they begin a new phase of life, having resolved their central difficulties and having made the necessary adaptions to their lives, having released once dormant potentialities. The midlife crisis is invariably a unique opportunity to grow and develop our personalities in a direction that will give our lives a deeper sense of meaning and purpose, a greater sense of fulfilment and a less troubled, richer engagement with our own true natures and the world around us.

A hundred years ago Freud and Jung produced two works that were to have an immeasurable impact upon the theory and clinical practice of psychotherapy. In 1920 Freud published *Beyond the Pleasure Principle*, which gave a comprehensive account of his repetition theory, while in the same year Jung was writing *Psychological Types*, in which he described for the first time his concept of 'individuation'. These thoughts on the psychology of midlife have running through them continual references to these two theories which stand as cornerstones of the psychoanalytic revolution, two different models of the human psyche and two contrasting approaches to the practice of psychotherapy. Both, however, shed light on why this psychological rite of passage not only appears to be a necessary experience in our emotional development but also the role of the midlife crisis as a significant part of our evolution as a species.

There are perhaps three evolutionary processes that sculpt humanity's development – biological, technical and ethical evolution. Biological and technological evolution are both accepted mechanisms of human development, but I would argue that ethical evolution provides a third tier to humanity's progress which becomes essential if we are to control our species' technical advance, which if left unrestrained will put our existence at risk.

If our ethical progress does not keep up with our technological progress, the planet and our species will be placed under perpetual threat. Our midlife experience therefore is crucial in providing our species with the wisdom, compassion and altruism necessary to guide humanity safely through the challenges that lie ahead.

– Eldership, the Cuban Missile Crisis and Jung's Breakdown –

Only two species of mammal have a post-reproductive life that lasts longer than their reproductive life. The first of these is the killer whale, or Orca. Pods of these whales are often led by very old females whose long experience and deep intuitive knowledge allow them to search out the rich grounds of food necessary to keep the pods well fed. The other species of mammal is *Homo sapiens.*

While the evolutionary purpose of the killer whale's many years of extended life so far beyond their reproductive years seems clear, what of the Orca's human equivalent? One of the very few psychologists who has questioned why human life extends so far beyond the child-rearing years was Carl Jung.

Jung found an answer in the observation of primitive tribal cultures where elders are the protectors of the ethical and cultural wisdom of the tribe and it is this eldership which preserves the cultural heritage and the moral integrity of the community. For Jung, if a culture is to maintain its deepest, profoundest roots, while moving forward to embrace the challenges of historical and technical change, it needs to find a balance between the energy, vigour and competitiveness of those in the first half of life and the experience, dignity and wisdom of those in the second. The young braves or warriors of the tribe, with all their zest and vigour, ultimately obey the rules of the elders, and this equilibrium provides a culture with the essential balance between audacity and prudence, impetuosity and foresight, energy and moderation.

A striking example of how the wisdom of elders can constrain the belligerence of tribal warriors took place in October 1962 when humanity faced what the historian Arthur Schlesinger called 'the most dangerous moment in human history.' On the morning of 16 October, the director of the CIA presented President Kennedy with irrefutable evidence that the Soviet Union was installing nuclear missiles on the island of Cuba, ninety miles from the US mainland. By mid-morning Kennedy had convened a meeting with his military chiefs, by the end of which the President had all but decided upon an immediate air strike against Cuba followed by a full-scale invasion of the island. Kennedy then was scheduled to have lunch with his US Ambassador to the United Nations, the veteran American politician and diplomat Adlai Stevenson. Kennedy greatly prized Stevenson's experience and wisdom and had publicly stated that 'the integrity and credibility of Adlai Stevenson constitutes one of our greatest national assets.'

Kennedy invited Stevenson into the Oval Office, explained his predicament and told him that he was about to activate the full military option advised by his generals. Stevenson was deeply concerned to hear about the Soviet actions, but was even more horrified to hear about the massive military response that Kennedy was about to unleash and insisted that there should be no air strike or invasion until every possible peaceful solution had been explored. He also advised that the motives behind the Soviet's reckless strategy should be carefully examined before responding with an equally reckless military approach. Kennedy then discussed the veteran Stevenson's warning with his brother Bobby and thankfully the US military action was postponed.

For the next thirteen days two conflicting views polarised around Stevenson's diplomatic, less bellicose approach and the aggressive, belligerent views of the military leadership led by the head of the army, Maxwell Taylor, and the head of the air force, General Curtis LeMay. LeMay was a particularly unpleasant individual who regarded his greatest achievement as 'Operation Meetinghouse', an air raid which took place in March 1945 when 325 B-29 bombers incinerated sixteen square miles of Tokyo, killing 100,000 civilians. It was the most deadly, most violent four hours in human history. The power of LeMay's invective now seemed to be propelling the argument towards a nuclear strike against Cuba with consideration given to a pre-emptive strike against the Soviet Union itself. Yet as the debate continued, the doves, led by Stevenson, slowly began to prevail over the hawks, led by LeMay.

Finally it was Stevenson's advice that Kennedy followed and the President chose a naval blockade of Cuba rather than the military strategy of bombing or invasion. The President also implemented Stevenson's proposal that the US should offer to exchange their missiles based in Turkey for the Cuban based Soviet missiles. The Generals were, of course, violently opposed to this plan and regarded it as a sign of weakness, but this gesture of reconciliation became US government policy which greatly reduced superpower tensions. Once the naval blockade was imposed Stevenson was also significantly involved in the formulation of all communications that Kennedy sent Russian premier Nikita Khrushchev until the crisis was resolved. At this critical moment in history – the nearest we have come to species extinction – a kind of natural selection was at work. Subsequent historical analyses suggest that had Kennedy taken the bombing and invasion option, there was a high probability that catastrophic nuclear conflict would have followed. The species was saved, however, by the prudent counsel of a wise elder which produced a measured, more compromise-orientated approach: a policy based upon communication and understanding in all matters regarding nuclear weapons.

Three years before the Cuban Missile Crisis Jung, in his famous BBC 'Face to Face' interview, was asked by John Freeman: 'Looking at the world today, do you feel that a third world war is likely?' Jung answered:

We are so full of apprehension and fear that one doesn't know exactly what it points to. One thing is sure. We need more psychology. We need more understanding of human nature, because the only real danger that exists is man himself. He is the great danger, and we are pitifully unaware of it. We know nothing of man, far too little. His psyche should be studied because we are the origin of all coming evil.

Jung's remarks forewarned humanity about the threat of nuclear war, so narrowly avoided in 1962. They also seem prophetic in regard to global warming and our current ecological crisis, for the 'evil' he refers to is surely a reference to rampant, chauvinistic narcissism such as was exemplified by Curtis LeMay and more recently was so blatantly on display as we watched the crazed antics of Donald Trump. Jung's insistence for the need for far 'more understanding of human nature' implicitly suggests we must involve ourselves in much more self-examination and self-scrutiny. Jung also makes clear that only with a new psychological approach can we overcome the dangers that we now face as a species, with far more emphasis on a developmental curve towards individuation.

From 1921, when Jung first introduced his theory of individuation in *Psychological Types*, until his death in 1961 he continually referred to – and redefined – this concept, which maintains that every individual is programmed to have an innate developmental drive that strives to release as much of the potential which lies latent in each of us towards maximum expression. The purpose of life, then, becomes this compulsion to reveal and extend the finer parts of our nature. This urge is facilitated by a succession of challenging growth points, or stages in life, which we may or may not take advantage of, depending on whether our complex or neurosis blocks this developmental impulse. The psyche's prime purpose, for Jung, is the thrust towards self-realisation based on the premise that we exist in order to develop. Yet this development can only be achieved through much exacting and painful experience, including the key phase in our middle years that we've come to call the midlife crisis. Individuation can prise open a complete range of life-changing personal qualities. It can transform how we conduct our closest relationships; it can deepen our compassion as our self-centredness diminishes; it can replace our narcissism with empathy; it can amplify our capacity for humility; it can soothe our concerns about our mortality. And yet this is only possible if we acknowledge and confront our demons – the fears and anxieties that rise up from our complexes and neuroses and our deeper, even more intimidating shadow. It constantly requires us to stray beyond our comfort zones, out into realms of being where we are exposed and vulnerable. But it's not just the individual whose life changes: the effects on the evolution of our society and culture can also be significant.

The case of the Cuban Missile Crisis is, of course, a very extreme example of the critical presence of individuation at a moment of great peril, but modern history gives us a number of examples of how the presence of unusually developed individuals have played a crucial role at key moments when a country or a culture has faced an especially testing challenge. In all these cases the rare qualities of the individuals concerned were forged by their exacting midlife experiences that produced a resilience, an empathy and an integrity that enabled them to extend the moral and ethical values of the cultures in which they lived at a time of historic crisis.

Nelson Mandela's pivotal role when South Africa's policy of apartheid was replaced by democracy and the country's first Black government comes to mind, for it was Mandela's reputation and example as a beloved, wise elder that steered South Africa through those perilous days, avoiding the predicted revenge bloodletting. Another such case occurred in 1947, immediately after Indian independence when mass outbreaks of murderous hatred between Muslims and Hindus were brought to a halt by the intervention of Mahatma Gandhi, whose authority and reputation were powerful enough to bring the violence to a close. Two of the greatest American presidents, Abraham Lincoln and Franklin Roosevelt, who led their country through the three greatest crises in US history – the American Civil War, the Great Depression and the Second World War – both suffered gruelling ordeals in their middle years which most people would have been overwhelmed by, but which transformed them into figures of immense historic importance who significantly contributed to the development of Western civilisation.

For Jung true individuated eldership, of the type displayed by Stevenson, Lincoln, Roosevelt, Gandhi and Mandela, can only be forged on what he calls 'the anvil of crisis', the rite of passage that Jung and the post-Jungians have described as the 'midlife transition' which is more widely referred to as the midlife crisis. Jung presents it almost like some Darwinian process of natural selection, whereby only those who brave the challenges and turmoil of the midlife transition successfully have what is psychologically required for wise, individuated eldership. Indeed all five of these 'elders' experienced significant midlife crises that had a transformatory effect on them. We will explore Lincoln and Roosevelt's middle years in much greater detail, but perhaps it is Jung's experience that we should first consider for Jung was not only a theorist of the midlife transition but also experienced four gruelling years of psychological breakdown which began in 1913 and continued until the end of 1917. In many ways his experience is the prototype of the midlife crisis, containing so many of its aspects and features which he was to write about so perceptively in the years that followed. To understand its origins, we must go back to the beginning of Jung's life, which he gives an account of in his autobiography *Memories, Dreams, Reflections*, written at the very end of his life.

'One memory', Jung writes in chapter one, 'comes up which is perhaps the earliest of my life. I am lying in a pram in the shadow of a tree. It is a fine warm summer day, the sky blue and golden sunshine is darting through green leaves. I have just awakened to the glorious beauty of the day and have a sense of indescribable well-being. Everything is wholly wonderful, colourful and splendid.'

In my own clinical experience, I always ask my clients about their earliest memories. In this, I am taking Freud's advice who places considerable significance on them: 'The recollection to which the patient gives precedence, with which he introduces the story of his life is the very one that holds the key to the secret pages of his mind.' Mostly these memories are populated by mothers, fathers and siblings and are often very instructive, and yet Jung remains alone in his earliest recollections but seems serene and content in his solitude. Indeed, as soon as people appear, things go awry. 'My mother holds me back and sternly forbids me to go into the garden.' Or more significantly:

> My mother spent several months in a hospital in Basel and presumably her illness had something to do with the difficulty in the marriage. I was deeply troubled by my mother's being away. From then on, I always felt mistrustful when the word 'love' was spoken. The feeling I associated with 'woman' was for a long time that of innate unreliability.

This absence of any parental presence suggests difficulties in bonding and forming attachments with both Jung's mother and father. Indeed, the prevailing early memories of his mother appear to be of her absence and the lack of trust and love that Jung felt for her as a consequence. His first memory of his father seems no better: 'I am restive, feverish, unable to sleep. My father carries me in his arms, pacing up and down.' It seems the world of interaction with his parents is filled with insecurity and unreliability and this first psychological wound marks the earliest origins of a process that would slowly gather a critical mass which would eventually tip him into his midlife crisis. Meanwhile to compensate for the lack of a secure, emotional bond with either parent, the infant Jung appears to have turned to the inanimate world of nature to find a soothing sense of security. Familiar sounds, shapes and smells provide a consistent sense of comfort. They provide stability and a feeling of safety and protection that the parental presence lacks.

At the age of eleven, while walking back from school, Jung had a defining, almost visionary experience, that he returns to again and again:

> Suddenly in a single moment I had the overwhelming impression of having just emerged from a dense cloud. I knew all at once: now I am myself. At this moment I came upon myself. Previously I had existed, but everything had merely happened to me. Now I happened to myself.

After this moment of heightened awareness Jung realises that 'I was actually two different persons.' Personality number one was engaged in the outer world, a tangible, known person that could succeed at school, at university and in his chosen profession. Personality number two was mysterious, unpredictable, enigmatic. When with his number two personality: 'I was my true self. As soon as I was alone, I could pass over into this state. I therefore sought peace and solitude in this "other", this personality number two. The play and counterplay between personalities number one and number two has run through my whole life.'

This duality was to be the central dynamic in Jung's psychobiography and is given external expression in so many aspects of his life. His need for what amounted to two marriages (to Emma Jung and Toni Wolff); his need for two homes (the family home at Küsnacht and his secret tower at Bollingen); his life as a public figure of growing international fame and his solitary creative process that produced the achievement of his vast collected works. Jung later came to describe these two states of being as 'extrovert' and 'introvert', terms he bequeathed to our contemporary lexicon. Indeed as he emerged from adolescence he was to experience fifteen years of intense extroversion.

From 1895 to 1900 Jung attended the University of Basel where he studied natural sciences and medicine. After qualifying as a doctor, he specialised in psychiatry and moved to Zürich to study and work at the Burghölzli Mental Hospital where he formed a close relationship with the clinic's director Eugen Bleuler, who promoted him to his deputy. He was then appointed Lecturer in Psychiatry at the University of Zürich, where his lectures became so popular that they had to be moved to a much larger hall. Jung's papers and articles began to attract international attention and by 1906 Freud had become aware of the brilliant young psychiatrist. By mid-1906 the two men began a regular correspondence, which was followed by their famous meeting in Vienna on 3 March 1907. Both men were well known for their magnetic charm and on this occasion, they made a seismic impact on each other. It was instant infatuation. At this historic Sunday lunch, their wives and Freud's children sat in silence as the two men, ignoring everyone else, plunged into a profound conversation about their work. Martin, Freud's son, wrote of the occasion: 'Jung did most of the talking and father listened with unconcealed delight.' Their conversation went on for thirteen hours, finishing at two in the morning.

This first encounter was followed by a week of further meetings and then by their rather touching correspondence. Jung ended his first letter to Freud: 'I hope my work for your cause will show you the depths of my gratitude and veneration. I hope and even dream that we may welcome you to Zürich. A visit from you would be seventh heaven for me personally.' Freud replied to Jung: 'Your visit was most delightful and gratifying.

You have inspired me with confidence in the future. I could hope for no one better than yourself, to continue and complete my work.' Very soon Freud was describing Jung as his 'Crown Prince', his future successor.

In 1899 Jung had met Emma Rauschenbach, the daughter of one of Switzerland's richest industrialists, when he visited her family home as a young doctor. He was immediately smitten by the seventeen-year-old and the couple married in 1903 after which they moved into an apartment in the Burghölzli which was quickly filled with the arrival of three children during the next five years. However, the harmony of Jung's family life was to be interrupted towards the end of 1905 when he began to treat a young Russian called Sabina Spielrein who he fell in love with. Jung wrote thirty-four letters to her which are extraordinarily passionate. 'It is my misfortune that my life means nothing to me without the joy of love, of tempestuous eternally changing love,' he wrote. 'I am looking for a person who can love without punishing, imprisoning and draining the other person. A love that is boundless and without reservations.'

The love affair that now developed was revealed to Freud by Jung in a letter dated 4 June 1909 and is repeatedly discussed in their subsequent correspondence as the younger man sought advice and support from his older mentor. Emma, spurred on by her sister Gret, was considering divorcing Jung as she had to endure 'the humiliation and social embarrassment' that the gossip surrounding Jung's relationship with Spielrein was causing her. With Gret's encouragement, Emma presented her husband with an ultimatum: he must end his relationship with Spielrein, leave the Burghölzli and either find a house for their growing family or else have one built. If he refused any of these conditions, she would divorce him. Freud, who liked and admired Emma, was firm in his advice to Jung who accepted all three of his wife's conditions. In 1908 he ended his association with Spielrein and he agreed to leave the Burghölzli. In the spring of 1908 Jung and Emma, on a Sunday afternoon walk, found a plot of land for sale on the shores of Lake Zürich in the nearby village of Küsnacht where they were to live for the rest of their lives.

During the decade between 1903 and 1913, Jung lived an extremely eventful life full of extroverted activity and outward achievement. Professional success at the Burghölzli was followed by academic success in Zürich and his international reputation was established as his partnership with Freud prospered. He lectured widely throughout Europe and the US. His private life had also proved eventful. The penniless young psychiatrist had in ten years become a kind of superstar in international medical circles.

Jung was a great admirer of the Greek philosopher Heraclitus and was particularly drawn to his concept of 'enantiodromia', a tendency in all systems and in all human experience whereby the more extreme a position that the system or person adopts, the greater the tendency for the system or individual to experience a complete swing to the opposite extreme. For instance, if a pendulum is held up to the horizontal axis when it is released, the gravitational pull will swing the pendulum over to a similar axis on the opposite side of the pendulum arc. Jung noticed how this propensity occurred regularly in the lives of his patients and he also observed how it appeared as a compensatory feature within families, particularly from one generation to another. Indeed this phenomenon was significantly present in his own family. His father, an impoverished rural pastor in the Swiss Reformed Church, was strickened by a kind of spiritual lethargy, continuing to perform religious rituals he no longer believed in. Jung, on the other hand, was spiritually adventurous. While his father remained a victim of depression and psychological inertia, Jung was committed to an almost fanatical pursuit of inner development. He believed that this enantiodromia reached its most intense presence during the midlife transition. As Jung was later to write of his midlife crisis: 'At the stroke of noon the descent begins. And the descent means the reversal of all the ideals and values that were cherished in the morning.'

Jung's own dramatic descent began at the beginning of 1913 when he wrote his letter of bitter farewell to Freud, permanently ending the relationship that had been so crucial to him. Later in the year not only did he resign as President of the International

Congress of Psychoanalysis but he also retired from Zürich University, giving up his position as Professor of Psychiatry even though he was only thirty-eight. It seemed his life of extroversion was changing as he was drawn increasingly into an introverted world of intense self-examination. He also gave up his international lecturing career. His sense of well-being deteriorated further because of a new relationship with yet another patient, placing an immense strain on his marriage and family life. Toni Wolff, a young woman of twenty-two, had arrived at Küsnacht for treatment with Jung in the summer of 1910 in a state of inconsolable grief which had tipped into clinical depression following the death of her beloved father. Within the course of the next two years, Wolff had changed from being Jung's patient to being his assistant and collaborator. Then in September 1913, after Emma had become pregnant with their fifth child, Jung suggested to Wolff that they take a holiday together to Ravenna, one of his favourite cities. It seems likely that it was on this trip that they became lovers. As the year progressed every afternoon Wolff would visit the Küsnacht house to work with Jung. It was clear to everyone what had happened and the impact on Emma was devastating. To escape this latest humiliation, Emma used her last pregnancy as an excuse to become a virtual recluse, barely seeing anyone other than her children. It was almost as though throughout 1913 Jung was attempting to deconstruct his life in an act of emotional vandalism.

As 1913 moved into December, Jung's mood darkened further and on the night of the eighteenth he had a dream that he regarded as a gateway into a new phase of life which seemed to be pushing him close to the edge of insanity. In this dream he was accompanied by a strange, mysterious figure whom he described as a 'savage' and together they were waiting, just before dawn, in a mountain landscape. As the sunrise began to fill the sky, the hero Siegfried's horn sounded across the mountains and Jung was filled with the dreadful realisation that he had to kill Siegfried. He and the savage were armed with rifles and lay in wait, ready to ambush the great hero. Siegfried then appeared on the crest of the mountain, which lay before them. Mounted on a chariot made of human bones, he drove towards them and just at the right moment Jung and the savage opened fire. Jung turned to flee, filled with disgust and remorse for having murdered such a beautiful young man.

Jung felt that this dream was so significant that an inner voice gave him a grim instruction: 'If you do not understand the dream, you must shoot yourself.' Under pressure from this menacing command, Jung explored the dream's meaning. The hero Siegfried was a role he had taken up for himself during the previous decade. 'The dream showed that the attitude embodied by Siegfried, the hero, no longer suited me. Therefore it had to be killed . . . This identity with my heroic idealism had to be abandoned, for there are higher things than the ego's will, and to these one must bow.'

So Siegfried's death, in the dream, signalled the end of his previous life and the beginning of something new, but before this new identity or new orientation could appear, Jung's old ego had to die. In the next few years he was often assailed by thoughts of suicide, but, as he said, these thoughts were the manifestation of 'egocide'. Jung now entered that period of confusion and turmoil that psychotherapists describe as 'liminality', a state of being between two positions, a period of waiting, of limbo, of purgatory, while the psyche is reorienting. For anyone facing this challenge, it is a frightening period of change when nothing is certain: when the sufferer is surrounded by fears and anxieties which feed on the old egocentric convictions of the past order that have now lost their relevancy. 'I stood helpless before an alien world,' Jung wrote, 'everything in it seemed difficult and incomprehensible. I was living in a constant state of tension.' Beset by a whole series of fantasies, dreams and visions, he felt he must engage with these unsettling images, which he increasingly wrote down.

In these descents into his unconscious he met a series of archetypal and mythological figures who became inner companions in this exploration of the deepest recesses of his psyche, including Elijah and Salome, Helen of Troy, Klingsor and Kundry. But the figure that Jung most welcomed was Philemon who was destined to accompany Jung, throughout his life, as an expression of an individuated wise elder.

In an attempt to place some sort of order on his chaotic state of mind, Jung developed a daily routine which he rigorously maintained. The first hours of the morning would be spent in his study recording his dreams and giving an account of his ever-shifting fantasies with all their archetypal figures. These writings were recorded in his 'black book', which then were transferred into his 'red book', with an accompanying commentary. He would then spend the second half of the morning with Wolff who became a very effective therapist for Jung, as she was the one person who seemed to understand the torment he was going through. Emma magnanimously later wrote of this period: 'I shall always be grateful to her for doing for my husband what I or anyone else could not have done at that most critical time.'

Jung's afternoons were spent down by the edge of the lake where he would fashion whole villages with houses, churches and farmyards out of rocks, mud and small pieces of wood. This would become an obsessional daily activity and seemed to be the one thing that gave him pleasure. After an hour or two by the lake he would see a patient, as a small amount of clinical work seemed crucial, which was followed by a silent, often tense, dinner with Emma, after which he would retire to his study and end the day making further entries in the 'black book'.

Four years later, Jung eventually emerged from his crisis. 'The years when I was pursuing my inner images were the most important of my life – in them everything essential was decided. It all began then. The later details are only supplements and clarifications of the material that burst forth from the unconscious, that at first had swamped me. It was the 'prima materia' for my lifetime's work.'

This quote from *Memories, Dreams, Reflections* is very revealing for Jung was clearly stating his belief that the experiences he had been confronted with during his protracted breakdown provided the clinical material upon which he would base his future psychological theories. These profound yet harrowing years of midlife transformed him into the man whose wisdom and insight seem so inspirational to his many admirers today. In his description of his midlife ordeal, he gives us a perfect example of how, if we can embrace these uncertain and frightening experiences, we will emerge liberated, somehow cleansed of aspects of our nature that previously held us back and kept us tied to personality traits that were detrimental to us and our capacity for development.

– Freud's Repetition Theory and Jung's Concept of Individuation –

In 1906 the Geneva Convention ruled that prisoners of war who escaped from belligerent countries to Switzerland had to be secured in internment camps and the Swiss authorities, with their customary flair for order and efficiency, organised a variety of arrangements for over 10,000 British prisoners during the Great War. In September 1917 Jung, as part of his compulsory Swiss military service, was unexpectedly appointed Commandant of the Château d'Oex internment camp near Lausanne and this bizarre twist in his career turned out to be just the experience Jung needed to help him recover his equilibrium after the emotional turmoil and psychological collapse that he had suffered during the previous four years. Furthermore his job as Camp Commandant was far from onerous, giving him plenty of time to work on a manuscript that was to become *Psychological Types.*

On its publication in 1921, one of its first readers was Sigmund Freud. Freud and his circle had hoped to consign Jung into a lapsed footnote in the history of psychoanalysis and the news from Zürich was that the deposed crown prince was locked into a prolonged psychotic breakdown, which didn't surprise Freud and his colleagues. Therefore, the sudden appearance of this comprehensive, well-argued book was an unwelcome surprise to Freud, unsettling his gratifying sense of schadenfreude. Having read it, however, Freud was reassured and wrote to his future biographer Ernest Jones: '*Psychological Types* is the work of a snob and mystic, with no new ideas in it. No great harm to be expected from that quarter.'

In fact Freud regarded any dissent from his own psychoanalytic theory as tantamount to instability and deviancy, which would result in the frequent collapse of his closest partnerships. Freud was a man who throughout his life took offence easily and any kind of theoretical or clinical difference was seen by him as an act of betrayal. It is therefore the heaviest of ironies that the one book that challenged the primacy of sexuality in psychoanalytic theory was written in his own hand. As Freud poured scorn on Jung's return to professional credibility, he had just published *Beyond the Pleasure Principle* in 1920, which amazed his colleagues when it stated: 'Based upon observed behaviour in the transference and upon the life histories of men and women, we have to assume that there really does exist, in the mind, a compulsion to repeat that overrides the pleasure principle.' 'The pleasure principle' was Freud's euphemism for sexuality and here he was, in the very title of his latest book, questioning the centrality of sexuality. For those who had kept the faith, Freud's *volte-face* was an extraordinary turn of events. Hadn't Freud, year after year, taught them that the most essential element of their enterprise was the primacy of sexuality over all other propositions and that this should never be questioned? And yet now Freud had come up with a new theory – the theory of repetition – that he maintained had even more potency than sexual drive.

This presiding compulsion to repeat, as Freud maintains, is something that most therapists will observe constantly in their clinical practices. It appears that human beings have an obsessive desire to continue patterns of behaviour that are often injurious to them because the 'familiar' appears to be much more attractive than the 'curative' and the client's natural inclination is to preserve at all costs the familiar. This impulse stands at the centre of our complex – that neurotic, dysfunctional response we fall back into when faced with difficult situations – which triggers a regression into infantile reactions that have such a damaging presence in our lives. Freud writes:

> Patients repeat all of their unwanted situations . . . the impressions they give are of being pursued by a malignant fate or possessed by some extraneous power. But psychoanalysis has always taken the view that their fate is, for the most part, arranged by themselves and determined by early infantile influences.

In his book Freud gives four specific examples of this – the benefactor who is eventually abandoned in anger by all of his protegés; the man whose friendships invariably end in betrayal: the lover whose love affairs always end badly; the woman who marries three successive husbands, each of whom falls ill and whom she nurses on their death beds. This, then, appears as Freud's central vision of the human condition. At the centre of our nature is a desire to always return to the old patterns of our earliest experience, which according to Freud we are programmed to repeat and all psychotherapists can offer is to soothe the client as he or she is battered by this insistent repetition.

In my own practice, virtually every client has displayed this neurotic tendency to replicate patterns of behaviour which were first experienced in their childhood, amplified by the emotional culture of their original family. This is their complex that we are attempting to subdue and it has a particular influence on the partners that my clients tend to choose. One client always chose angry partners, replicating the anger of his mother. Another client, the youngest of seven children who received minimal attention from her exhausted parents, remained married to a man for twenty years who had Asperger syndrome and took little interest in her emotional welfare.

Jung's concept of individuation, lying at the centre of his view of human nature, is very different, insisting that at the core of our personality lies a set of innate potentialities that we are given a lifespan of up to ninety years to discover and express. At some stage in our lives, these potentialities may be liberated and become a central feature of our unique individuality. These fundamental aspects of our true nature are unconsciously present from our earliest days, rather like the genetic code in an acorn that has the capacity to transform its tiny form into a mature oak tree. And yet this transformation is impeded by the manner in which, during early infancy, we swathe ourselves with layers of protective defences as we meet the inevitable challenges, tensions and traumas of our earliest years. The extent of these protective measures will depend upon the level of threat and trauma that we experience and feel as infants and as they build up, they will smother and choke our true selves, using strategies such as denial, repression, regression and projection that will then deny us contact with – and the expression of – our authentic individuality. These defenses will also use the device of repetition to keep themselves embedded in each of us. Yet Jung insists that there is present in the human psyche an even more powerful mechanism than this urge to repeat: an innate drive which seeks to break out of the confines that our defences place us in. This is the beginning of the individuation process, which makes its first appearance in the experience that we have come to call the midlife crisis.

While Jung's critics will point to his idealistic naivety and his gullible optimism, Freud's critics will focus on his curmudgeonly pessimism, his constraint of the human spirit and his cynical despondency, triggered in 1920 by his despairing view of humanity after the horrors of the First World War. And yet I see something in the approach of these two great innovators that is complementary. It is almost as though we need both Freud's pragmatic left-brain realism and Jung's soulful right-brain intuition. It is as if these two contrary visions of human nature together make up a complete whole, the two halves between which we all oscillate. Indeed, clinical practice suggests that both points of view are present. On the one hand repetition theory seems embedded in our clients' experience, these repeated patterns of familiar, habitual experience blighting their lives. Like moths drawn to the flame they return again and again to predictable styles of behaviour that perpetually work to their disadvantage. This constant return to the familiar is the bedrock of neurosis, the originating feature of our complexes.

So for Freud, the therapist is there primarily to help her patients find a means of carrying life's burden in the most comfortable way possible as they continue their addiction for repetition, but the therapist cannot release them: she cannot cure them from their itch to replicate. Jung, however, provides us with a way out by suggesting that the very purpose of life is for us to move beyond the endless wheel of repetition into a state of being that takes us into a whole new realm of experience. In the Jungian model, psychotherapists are called upon to help their clients overcome the repeated patterns of neurotic behaviour that our defences insist we continue to replicate. The therapist's role then becomes that of a travelling companion, joining the client on this odyssey of inner exploration and self-discovery that the midlife phase of life demands of us.

Jung's optimism suggests a way of transcending Freud's pessimism. However, Freud was not wrong. He described with visionary precision the nature of our ailment, but he left us there, with no way out. Yet Jung provides us with a means of salvation. This suggests that these two competing drives are to be found in every human psyche. I would go so far as to say that resolving this tension between these two polarised compulsions, which are in a state of constant friction and discord, is one of the prime psychological challenges we are all faced with. No individual psyche is either solely the conservative preserver of the familiar or conversely the restless devotee of change and renewal. It is our fate always to be both; a universal tension within us all; a constant chaffing between our deep desire to remain the same and our unrelenting desire to develop and change.

If the compelling urge to repeat detrimental patterns of behaviour and the latent capacity to develop beyond these patterns both exist in the psyche, despite their seeming incompatibility, can we combine the Jung and Freud models to create a kind of unified theory? As has already been mentioned, but needs emphasising because of its importance, Freud's repetition theory is clearly a central element in all our complexes, but Jung proposes that within the psyche is the mechanism he calls individuation which seeks to staunch the impact of our compulsion to repeat and which allows us to make a psychological breakthrough that provides us with the means to escape the clutches of our complex.

Is Jung right? Is the most powerful element of the psyche its capacity to develop, to move forward, to construct a more complete, fulfilling, rewarding mode of being, breaking free from our ingrained complexes? Perhaps Jung's own life can answer these questions.

As we saw in the last chapter, in infancy Jung was deeply disturbed by his mother's frequent and prolonged absences, as she needed regular breaks from the tensions of her marriage. 'I was deeply troubled by my mother's being away,' he wrote. 'From then on, I always felt mistrustful when the word "love" was spoken.' This inner disturbance was repeated again and again in his adult life as he attempted to find, in his relationships with numerous different women, a means of recovering from his central emotional wound, which lay at the heart of his complex.

From 1905 when he began treating Sabina Spielrein until 1925, when his relationship with Toni Wolff began to wane, Jung's chaotic emotional life was filled with a succession of affairs and infatuations. Time and again between the ages of thirty and fifty Jung placed both his professional and personal life in jeopardy, causing himself and those closest to him considerable emotional pain because of his compulsive repeated need to be surrounded by adoring, devoted women. Emma had written to Freud in 1911: 'Of course the women are naturally all in love with him' and again and again she had to decide whether to accept his infidelities or divorce him.

This narcissistic need to be surrounded by women who loved him was a compensation for the absence of love that characterised Jung's relationship with his mother. From early 1908 until April 1912 Jung's affair with Spielrein had ebbed and flowed through a succession of emotional crescendos and diminuendos and during the autumn of 1910 he was having an affair with Maria Moltzer, a nurse at the Burghölzli clinic and possibly with another member of staff, Martha Boeddinghaus. Having promoted both these women to his assistants he wrote to Freud in a self-congratulatory tone: 'Between these two ladies there is naturally a loving jealousy over me.' All these romantic encounters ran parallel with Jung's developing relationship with Toni Wolff.

By September 1911, Jung's desire to flaunt these relationships in public reached its zenith when he insisted that Emma, Toni Wolff, Maria Moltzer, Martha Boeddinghaus and Sabina Spielrein all accompany him to the Third International Psychoanalytic

Congress in Weimar. He wrote proudly to Freud: 'This time the feminine element will have conspicuous representatives from Zürich: Sister Moltzer, Frl. Dr. Spielrein, then a new discovery of mine, Frl. Antonia Wolff and last but not least my wife.' Jung must have been gratified by Freud's reply: 'We Viennese have nothing to compare with the charming ladies you are bringing from Zürich.'

All the women listed by Jung attended the Congress except Sabina Spielrein, who called off at the last moment. Her absence from Weimar, and her subsequent move from Zürich to Vienna that autumn, seems likely to have been a reaction to her sense of rejection now that Toni Wolff was taking up so much of Jung's time. While Jung welcomed the developing distance between him and Spielrein, he didn't welcome the violent rows he was having with Emma who now suspected that her husband was having an affair with Wolff after another of Jung's spurned mistresses, possibly Maria Moltzer, tipped her off.

From their very first encounter, Jung recognised there was what he called 'a golden thread' linking him to his new patient; feeling that in Wolff he had found a woman with whom he could converse at 'the very deepest level of the psyche.' At last Jung felt he could fill that gap – that empty cavity in his nature – which had been hollowed out by his mother's neglect. Yet despite Jung's immediate attraction to Wolff, his professional integrity and clinical rectitude held up. After a series of analytical sessions, there appears to have been a period of no contact as Jung hesitated to pursue their relationship further, but the impact she had made on him would not relent and a series of dreams and synchronistic events confirmed his belief that his relationship to her was predestined.

While the intensity of Jung's feelings for Wolff was driven by his belief that she could heal his deepest wounds, there was an equally powerful desire that he remain committed to his marriage to Emma, because of her solidity and dependability. So acute was his need for both women that he insisted on a ménage à trois which required Wolff to make daily visits to their Küsnacht home. The impact of his behaviour caused misery throughout the household and these extreme tensions, combined with the end of his relationship with Freud, propelled Jung into his long breakdown, which would overwhelm him for the next four years. But the compulsions of his complex created an emotional chaos that would provide no answers to Jung's problems and salvation was not to be found in either Wolff's or Emma's arms and the crisis he had provoked did act as a starting point for a period of intense introversion and self-examination. Ultimately the answers Jung was looking for, the healing he craved, could only be achieved through his own efforts: by his own search for a state of being which would eventually give him the peace of mind and inner stillness that he longed for.

How then did Jung finally shed his repeated

compulsion to insist upon the devoted attention of two or more women who loved him? How did he break through his relentless hope that a retinue of adoring women would cure the psychic discomfort that pursued him? Having confronted all his demons and shadow material deep in his psyche, he was eventually to find his answer in a few acres of ground that he came across in the autumn of 1922, a few months prior to his mother's death.

When Jung returned from his stint as Commandant at the Château d'Oex internment camp, the extrovert side to his nature appeared to have been revitalised. Weekends were now often spent with his five children, sailing up the lake on his red-sailed yawl, on what he called their 'north-pole journeys'. He would sit in 'the admiral's seat' barking out orders to his crew. They often went to the upper reaches of Lake Zürich, where they set up camp, dug trenches for latrines and cooked food over campfires. These weekend adventures continued for several years, until late 1922 when Jung purchased a lakeside plot of land near the village of Bollingen. Soon after buying these four acres of ground, Jung's mother died suddenly at the age of seventy-five.

This unexpected event spurred Jung into action. As if in response to his mother's death, he felt he must build a stone tower on the land that he had just acquired. He drew up some initial designs and purchased suitable stones from a quarry on the opposite side of the lake. For the next year, he set about building his tower, which became the very centrepiece of his life. Nothing seemed more important. Each weekend, accompanied by his son Franz, and his son-in-law Kurt, he would sail down the lake and continue the construction work. He felt he was 'being reborn in stone.' It was as if he was discovering a part of himself that had always eluded him but which he intuitively knew was there, deep in his unconscious. At last, at the age of forty-seven, he felt this significant, once dormant, element of his nature was being released and the energy that the building of the tower seemed to generate was available to him during the week as he wrote furiously.

At this time he was primarily writing about individuation and the 'self' and it was as if the tower was an outward manifestation of the individuation process.

> The feeling of repose and renewal that I had in the tower was intense from the start. It represented for me the maternal hearth . . . from the beginning I felt the tower was in some way a place of maturation – a maternal womb or a maternal figure in which I could become what I was, what I am and will be. It gave me a feeling as if I were being reborn in stone. It is thus a concretisation of the Individuation process . . . At Bollingen I am in the midst of my true life, I am most deeply myself.

Over the entrance of the tower he carved the words SHRINE OF PHILEMON – REPENTANCE OF FAUST.

Philemon was the senex, the wise elder that had guided him through the darkest hours of his breakdown; Faust was the divided personality, the narcissistic nemesis that had caused such misery for so many, as he had compulsively repeated his pursuit of women to soothe his bruised psyche. Now at last the wiser, mature part of Jung's psyche, symbolised by Philemon, had prevailed. Repeated intrusions from his complex were now replaced by the developmental path of individuation.

Using Jung's life as a case study, it becomes clear that the formulation of his theory of individuation was as much lived out as it was written about. Whilst the concept gets its first airing in *Psychological Types*, the theory of individuation is elaborated upon throughout the 1920s when Jung has had time to dwell upon and consider in depth the entire experience of his breakdown, his recovery and the impact the building of the tower has had on his development. Through these experiences his true self had gained the upper hand over his ego and its complexes, creating a transformation, a restructuring of his psyche which allowed the more authentic, more judicious, more enlightened, more benevolent elements of his nature to percolate up through the strata of his psyche. Jung's life, his very example, seems to suggest that the process of individuation can transform our personalities, allowing us to escape from the tyranny of Freud's repetition theory.

– The Mother and the Amygdala –

In 1537 a young medical student, who one day would become known as 'the father of anatomy', arrived at the University of Padua. So talented was this twenty-four-year-old student from Brussels that on graduating he was offered the position of Professor of Surgery. But Andreas Vesalius was not only a medical genius, he was also a kind of anatomical showman and very soon he was giving public dissections to large audiences, using the corpses of dead criminals. These performances attracted spectators from all over Italy and his future students included Bartolomeo Maggi, surgeon to Pope Julius III, Realdo Colombo, Michelangelo's friend and doctor, and Gabriele Falloppio, discoverer of the fallopian tube. Another young man inspired by Vesalius was Julius Caesar Arantius, who was later to become professor of surgery and anatomy at Bologna University. Arantius developed a particular expertise in brain anatomy and while working on his dissecting table, he came across a small structure in the temporal horn of the lateral ventricle of the brain he was dissecting which caught his attention. It appeared to be a separate sub-organ of the brain and Arantius thought it looked like a pair of seahorses. He therefore called it the hippocampus, the Greek word for seahorse. Arantius had no idea what function the hippocampus provided and this knowledge had to wait for 400 years when MRI brain scans and comparative brain autopsies allowed us to increase our understanding of what role the hippocampus plays in the process of neurological regulation.

Today we know that this versatile component of our brain has a number of roles but it is especially designed to perform one critical task – to attempt to keep the disruptive, undisciplined, panic-stricken practices of the amygdala under control. The amygdala is a cluster of almond-shaped cells located near the base of the brain that regulates our responses to perceived threat. It has an ancient provenance that provided earliest man with an alarm system which activates a fight-or-flight response that is programmed to detect every sabre-toothed tiger, every poisonous spider, every venomous snake that potentially lurks around every corner. Once aroused, the amygdala releases a jet of high-octane adrenaline, courtesy of the adrenal glands, throughout our body which empowers us to run faster, jump higher and anticipate every danger that our perilous world faces us with. But this optimisation of our physical capacities to escape the perceived threat has a number of unpleasant consequences. The glut of adrenaline rarely gets used up fuelling physical exertion. In our now sedentary circumstances, the adrenaline remains in our bodies unused and converts into the toxic substance cortisol that has a detrimental impact on our serotonin and dopamine levels which provide us with feelings of confidence and contentment. As the adrenaline and cortisol do their corrosive work, anxiety levels increase, a sense of well-being diminishes and the amygdala's worldview of permanent danger prevails, often plunging the fearful individual into depression as serotonin and dopamine levels dwindle.

Because of these overactive practices of the amygdala, the brain developed the hippocampus, which can distinguish between real danger and other forms of perceived threat that are in fact benign. This struggle between the adrenaline releasing amygdala and the calming hippocampus is at its most intense in our early years and is one of the most decisive factors determining whether our future life is one of fulfilment and reward or whether it is marked by a surfeit of anxiety, dissatisfaction and depression. But the amygdala has a significant advantage as it is fully developed in the foetus's brain by the beginning of the eighth month of gestation, while the hippocampus takes the first three years of life to become fully effective. So the amygdala is alert and active, unimpeded by the hippocampus, during the traumatic moment of birth.

What exactly is happening in the infant's brain at the moment of birth has been the lifetime's work of the psychiatrist and neurologist Stanislav Grof. His groundbreaking book *Beyond the Brain: Birth,*

Death and Transcendence in Psychotherapy was based on thousands of psychotherapeutic hours in which his clients, using LSD, were able to extend their memories back to the very moment of their birth.

Grof's account of the birth process describes an experience of extreme trauma. The perfect equilibrium of serene intrauterine existence is quite suddenly violently disturbed, first by alarming chemical signals and then by fierce muscular contractions. Despite these savage uterine convulsions, the maternal cervix is still closed and the baby has no escape. The way is blocked. Without warning the connectedness of intrauterine life is ruptured, replaced by feelings of overwhelming anxiety and an awareness of an imminent threat. 'The first stage of delivery,' writes Grof, 'is the experience of no exit or hell. It involves a sense of being stuck, caged or trapped in a claustrophobic, nightmarish world, an experience of psychological and physical torture.'

During the moments before birth, excessive amounts of adrenalin are pumped into the baby's brain by the terrified amygdala. But then when all seems lost a speck of hope appears as the cervix dilates. There is now a slim chance of survival, of escape. The dilating cervix propels the baby down the birth canal, but life is still not assured as the amygdala continues to warn of pending annihilation. The baby can't face what is coming and is desperate to end the whole process, but it is powerless to stop the contractions. Quite suddenly the fear of imminent catastrophe is replaced by a blinding light, a sense of release, a confusing convulsion of sight, sound, sensation and smell. There is a tremendous sense of all the faculties and erogenous zones providing an end to the horror of constriction, of darkness, of confinement. The baby has escaped. Whatever this new confusing state is, it is infinitely better than what preceded it.

Gaining these insights from his patients, Grof maintains that birth is a momentous trauma, affecting all our lives. The baby has passed from 'an initial condition of undifferentiated unity' into a harrowing ordeal. This moment of near-death terror is followed by an unexpected 'redemptive liberation', restoring, if the birth goes well, a harmonious reunion with the mother. The baby's sense of reassuring connection is re-established as a stream of the love hormone oxytocin bonds mother and child. Yet this experience of being flung from embryonic tranquillity to the terror of imminent annihilation, back to a resumed fusion with the mother, is a trauma that will be etched onto the unconscious of every individual.

With all my clients, at some stage in their therapy, we will try and reconstruct the critical features of their very earliest weeks and sometimes we manage to return to the circumstances of their birth. Yet unlike Stanislav Grof, I don't use LSD for this important work. A picture of their earliest days can be built up using dreams, active imagination, hypnosis and in particular by gathering memories from grandparents, parents, older siblings, uncles and aunts and other family friends, where these are available. I also ask clients to bring photographs of themselves as tiny babies with their mothers, which can be very revealing. In this fashion we can get back to the first days and weeks of their lives.

I had one client called Mel whose mother suffered such serious internal injuries during her birth that she was rushed off to a nearby general hospital, leaving her newborn daughter behind in the maternity hospital. In our work together we managed to return to her memories of the incubator in which she spent her first three weeks of life and recreate her state of mind as a newborn baby.

As Mel's life had subsequently progressed, she had become a woman of immense self-sufficiency, so independent that she never relied on others. However, at thirty-seven she desperately wanted children, but because she never allowed herself to become emotionally vulnerable enough to sustain long-term relationships, she was unable to find a partner or husband with whom she could have a child. As a midlife crisis now overwhelmed her, she was filled with a deep sense of loneliness and was frightened that she might never have children. Her self-sufficiency had become her prison. During therapy, in our search for the origins of her self-reliance, we eventually returned to that cold, metal incubator, where close maternal contact, with its flow of vital oxytocin, was absent. A newborn baby desperately needs to experience, in its first hours of life, intimate physical contact with its mother and soothing, palliative oxytocin to help recover from the ordeal of birth. Mel felt sure that as a tiny baby, she had intuitively realised that she was on her own. Memories of this baby lying in that incubator somehow shifted something in her psyche. It was almost as though her compassionate feelings for this baby lying in that metal box opened up in her a capacity for vulnerability, softening her self-reliance. During our work together she began her first long-term relationship, got married and had a baby.

When going through my own midlife crisis the key feature of recovery was returning to an event which occurred in the seventh week of my life. The love of my mother's life was her father, whom she adored. He had always longed for a son and as she was an only child, when I was born she undoubtedly saw me as a gift she wanted to share with him. Soon after my birth she therefore decided to return to live with her parents for a prolonged stay so he could enjoy the company of his first grandson. As my own father, a naval officer, was spending much of his time at sea, this seemed to everyone an ideal arrangement. In my seventh week we set off for our visit to my grandparents. Two days after our arrival my grandfather, a seemingly fit fifty-three year-old, suddenly collapsed and died of a fatal heart attack. My mother's few weeks of happiness following my birth disintegrated into months of misery, confusion and disappointment. My grandfather's sudden, unexpected death and my mother's reaction to it was one of the defining moments of my life and returning to it in my late forties was perhaps the most significant feature of my midlife crisis. This event and my belated memory of it was, I'm sure, the reason I became a psychotherapist, and indeed why I came to write this book.

Given the ordeal of birth, the presence of the overactive amygdala and the absence of the reassuring hippocampus, the newborn baby's sense of security will depend upon the mother's ability to calm, soothe and bond with her baby. This in turn will depend upon the emotional environment the mother finds herself in. If she suffers from post-natal depression or if, like my mother, she has to cope with some unexpected bereavement, there will be a minimal oxytocin flow. In this way, one of her roles is to do the job of the emerging hippocampus, calming her baby and fending off the fears and perturbations of her infant's amygdala. All this is a tremendous demand to place onto the mother, who may be facing either the natural anxieties of having her first child or the considerable pressures placed on her by one or more other children, unsettled by the arrival of this newcomer. The situation is made even more critical by the fact that excessive exposure to early stress and high levels of adrenaline and cortisol inhibits hippocampal growth.

In her book *Why Love Matters*, Sue Gerhardt, a

psychotherapist and co-founder of the Oxford Parent Infant Project, describes how for a baby the first sources of pleasure are smell, touch and sound and in particular the sound of their parents' voices, their scent and the feel of their skin. She emphasises how in infancy parents' looks, facial expressions and smiles stimulate healthy brain growth and also promote the healthy development of the hippocampus. When babies look at their mothers and observe their dilated pupils, which expresses loving attachment, they are gratified and reassured. This suppresses amygdala activity and releases beta-endorphins, dopamine and oxytocin into the brain which helps new cerebral tissue development. Indeed, with these optimal circumstances the baby's brain will double in weight in the first year.

If this doesn't happen, Gerhardt warns that high levels of cortisol in the baby's brain will close down the hippocampus's receptors, making it less able to inhibit cortisol release from the amygdala, further impairing hippocampal enlargement. Gerhardt cites Harry Chugani's 2001 research on Romanian orphans, who having experienced no parental touch or maternal affection in their austere, poorly resourced institutions showed – through MRI data and serotonin and dopamine checks – major hippocampus damage and minimum presence of serotonin and dopamine. So this is what we face in our earliest years of life. On one side, we have the trigger-happy amygdala, trying to persuade us that we face a daunting and frightening world full of menace and life-threatening danger. This contrasts sharply with the view from the emerging hippocampus's vantage point, which regards the world as much more benign and benevolent.

In this critical neurological balancing act there is so much at stake. Will the amygdala or hippocampal worldview prevail in the earliest years, laying down a lifetime's emotional outlook? Will the prevailing attitude be the optimism of the hippocampus or the pessimism of the amygdala? Will anxiety or a sense of contentment be the predominant attitude to life? Unbeknownst to her, what a mother is doing, during these earliest years, is nurturing her child's hippocampus and developing its vital capacity so that when she has to move on to other inevitable concerns and responsibilities, the hippocampus is developed enough to take over her calming, reassuring role.

Freud, Jung, Melanie Klein, Donald Winnicott and the other early pioneers of psychotherapy had none of the hard neurological data that is available to us today thanks to MRI scan evidence and the findings of comparative brain autopsies. Instead their theoretical conclusions were speculative, relying solely on the anecdotal evidence of their clinical work and the presenting symptoms of their patients. But now we know that these speculations, these theoretical models, were correct. The hard evidence confirms their theories. Indeed they are completely vindicated in spending so much time considering the crucial field of early infantile trauma. Their preoccupation with the emotional experience of early infancy and childhood and its impact on adult life was prescient and far-sighted and although they had no way of knowing of the vital tension between the amygdala and hippocampus, they were the first generation of medical scientists to scrutinise, study and catalogue the consequences of these neurological processes.

Of all the post-Freudian theorists who wrote on the subject of early trauma and infantile fear of abandonment, Winnicott's account is one of the most compelling. Winnicott maintains that the helpless infant senses both her powerlessness and her deep vulnerability as well as the essential life-giving presence of her mother. This indispensable maternal love Winnicott called 'primary maternal preoccupation'. If this maternal care is withdrawn or damaged the child will fall into an adrenaline drenched fear of abandonment. Yet almost inevitably this maternal attention will, at some point, be withdrawn, even if it is simply to attend to another child, triggering what Winnicott calls 'primitive agonies', with all the attendant fears and emotional confusion.

For Winnicott, the infant will attempt to steer around these primitive agonies by using defenses such as denial, deflection, projection and repression which are refined and developed in such a way to help avoid the full trauma of the feared abandonment. As the child grows up these primitive agonies then lie dormant in the unconscious, occasionally sending a spasm of anxiety up through the defensive layers of the psyche, causing symptoms ranging from containable depression to full-blown states of mental breakdown. Winnicott introduces us to his concept of primitive agonies in a paper called 'Fear of Breakdown', first published in 1974. In his view the word 'anxiety' is not strong enough to convey the infants' agony as they realise their complete powerlessness and the consequent dread of abandonment. In this state of absolute dependence Winnicott describes the primitive agonies as 'a return to an unintegrated state', 'falling forever', 'loss of the sense of the real', and 'loss of the capacity to relate'.

What Winnicott is suggesting is that if we are ever to rid ourselves of the spasms of anxiety and depression that will, from time to time, rise up from deep within our unconscious, we have to also return to the full experience of these infantile primitive agonies. He suggests we can do this by stripping away the defenses that we have wrapped around our early traumas. We can then re-experience these primitive agonies once again in midlife with the hope that by acknowledging and facing them, rather than deflecting or denying them, we can explore their nature, understand their origin, soothe their impact and find the resources which will help us overcome and integrate their corrosive presence.

Winnicott also maintains that we will only achieve this in our middle years when we have surrounded ourselves with sufficient emotional assets to assist us in this endeavour. Once these resources are in place, our unconscious realises that we are then actually ready to face the frightening reality of our primitive agonies full on as we unconsciously constellate a series of outer experiences that tip us into our midlife crisis. What the caring mother or the underdeveloped hippocampus was unable to provide, now the midlife adult must do for him or herself. The midlife crisis, therefore, offers the individual the opportunity to overcome and repair the damage the amygdala has done during birth and in early infancy and recover from the emotional frailty that the mother was unable to prevent, as she found herself having to attend to other competing responsibilities. But the individual concerned will not take on this demanding rite of passage until he or she has unconsciously put the necessary support in place. This could be an empathetic therapist, a compassionate spouse or family, a flexible professional arrangement, financial security or just a sufficiently developed temperament capable of withstanding the storms ahead. Jung was surrounded by all these things before he plunged into his midlife breakdown in 1913.

Just as Winnicott insists that our psychological and emotional difficulties will never be resolved 'unless the bottom of the trough has been reached, unless the thing most feared has been experienced', so Jung stresses repeatedly the need and significance of the period of turmoil in our middle years, when our fears of abandonment originating in our birth and earliest years rise to the surface of our psyche. During his breakdown Jung conceived a new vision, which we can all now benefit from: a psychological model that shows how Winnicott's primitive agonies can – during the midlife crisis – be finally shed and laid to rest.

In his new theory Jung insists that lying at the centre of our psyche is what he calls the 'Self', a pristine, undamaged authentic integrity which expresses the finer parts of our personality, or what Abraham Lincoln called 'the better angels of our nature'. This Self stands in contrast to our narcissistic, defended and troubled ego, which is constantly harried by the echoes of our primitive agonies. Jung suggests that if we are to develop psychologically we need much more contact with our self and while our ego is preventing this contact, we need to experience, during our middle years, what he refers to as 'egocide'. This egocide becomes the doorway into part of our psyche where the self is to be found. The midlife crisis, using Jung's language, forces us to make the transition from the 'ego-orientated life' to the 'self-orientated life'. In one of his final books, *Mysterium Coniunctionis*, Jung writes: 'The Self, in its efforts at self-realisation reaches out beyond the ego personality . . . The experience of Self is always a defeat for the ego.' It is this defeat of the ego and this elevation of the self that the midlife crisis is attempting to achieve. Psychologically this means that during the midlife crisis the ego exhausts its own resources and a space is cleared for the Self to fill the gap. Once the Self is fully engaged, a feeling of relief and renewal will emerge as a new healing energy is provided, bringing to an end the impact of the primitive agonies.

In his 2010 book *The Neuroscience of Psychotherapy*, Louis Cozolino describes the great plasticity and late growth potential of adult hippocampi and claims that effective, successful psychotherapy can facilitate hippocampal growth in later life: 'Psychotherapists are applied neuroscientists who create individually tailored, enriched learning environments designed to enhance brain functioning and . . . to promote neural network growth.'

It's an interesting parallel to Jung's model as what Cozolino means is that the mother's impossibly demanding job can be completed by the psychotherapist. Yet this belated psychotherapeutic cultivation of the hippocampus is often prompted when the individual concerned seeks out therapeutic help during the midlife crisis.

– The Eye of the Storm –

My mother had her midlife crisis at the age of eighty-six.

Quite suddenly her famous toughness and robust stoicism deserted her and for several months she sunk into an uncharacteristic depression. Then, for the first time in her life, she started taking antidepressants and fixed up to see a psychotherapist. Soon after, she rallied and appeared to become altogether different. Her no-nonsense self-sufficiency melted away and a softer, more sympathetic, more uncertain character emerged. This unexpected vulnerability allowed so much more intimacy between us and for the last four years of her life I became close to her for the first time. As a result, we were finally able to lay to rest all the tensions and friction that had been the hallmark of our relationship for sixty years and which no doubt had forced me into finding relief in decades of therapy.

How can someone have a midlife crisis at eighty six?

Very few of us get through life without facing some kind of existential challenge and because of our very defended natures these periods of turmoil can be delayed until the end of our lives when our defences become ramshackled and far less effective. This is why I am so sure that my mother's midlife crisis occurred so belatedly: because she was so well defended, so invulnerable, she had managed to reach this advanced age without experiencing anything so psychologically and emotionally demanding. Until then she remained, as it were, crisis-free with her stiff upper lip firmly in place, emotional qualities particularly valued by her generation.

In stark contrast to my mother Beethoven suffered his midlife crisis at the early age of thirty-two. Desperate to come to terms with his increasing deafness and trying to find a remedy to staunch its impact, he had a prolonged stay in 1802 in the small town of Heiligenstadt, a few miles to the north of Vienna, on the advice of his doctor. But this period of supposed recuperation plunged him into a state of acute despair that pushed him towards thoughts of suicide. We have a record of his state of mind in the form of a letter to his two brothers which he never sent, but which was found amongst his papers, after his death.

For six years now I have become increasingly afflicted and I'm finally compelled to face the prospect of this lasting malady . . . I have been compelled to withdraw myself, to live life alone as it has been impossible for me to say to people, 'Speak louder, shout, for I am deaf'. How could I possibly admit an infirmity in the one faculty which ought to be more perfect to me than others, a faculty which I want to possess in the highest perfection, a perfection such as few in my profession have ever enjoyed. For me there can be no relaxation with my fellow men, no refined conversation, no mutual exchange of ideas. I must live alone, like an exile.

The 'Heiligenstadt Testament' is partly a suicide note, partly an artistic mission statement, partly a cathartic exorcism and partly an affirmation of intent that despite his deafness he will achieve what he is certain he is capable of. It is a revelatory moment of crystalline clarification which, arguably, saved his life. As he sees his future with all its challenges, he accepts it and returns to Vienna where he composes some of the most remarkable music ever written. Without doubt, this was the moment of his midlife crisis which not only forged in him a means of overcoming his deafness, but also inspired the writing of his 'Eroica' Symphony, a work that launched the Romantic period that dominated European culture for the next one hundred years.

Using my mother and Beethoven as two unlikely bookends within which the midlife crisis can occur, I would suggest that this experience of profound transformation can happen at almost any time in adult life and is not defined by its timing, but must have as its one presiding element the sheer power of its transformatory impact.

I have noticed that my clients have a striking propensity for self-sabotage as they drift anxiously towards their own middle years and this 'ego deconstruction' is the first stage of the midlife crisis. As we have seen, the first half of life is governed by the drive and propulsion of the ego, with all its narcissistic power. The tenacity, flair and ambition of the ego is a necessary resource that provides the energy needed for professional advancement and the quest for a partner and family, and yet this narcissism is an inevitable corruption of the true authentic nature of the individual as it adjusts and modulates itself to fit into the demands of the culture and social norms it is surrounded by. It is this reconnection to the true self, that centre of potential wisdom in each of us, that is one of the essential features of the midlife crisis.

Jung's psychobiography is a clear example of this process of ego deconstruction. As he ended his relationship with Freud, resigned as President of the International Psychoanalytic Congress, retired from his professorship at Zürich university and came close to wrecking his marriage to Emma by starting his affair with Toni Wolff, one can see this process of ego deconstruction at work. Similar to Jung's experience I have had many clients who have left a marriage, resigned from a job, changed careers or moved far from a previous home as they enter their midlife ordeal. It's as though this wise self, in its struggle with the ego, has to facilitate this necessary crisis propelling the individual into a new psychological and emotional territory in the knowledge that a period of introspection and self-examination is an essential need at this stage of life.

The Chinese have a particular understanding of the possibilities that can emerge from a crisis, as their pictogram representing 'crisis' also signifies 'opportunity'.

The first stage in the midlife transition, ego deconstruction, is followed by the second stage – the experience of enantiodromia, which we considered in chapter two. Its appearance in the midlife crisis has two main elements: firstly, the ego's extroversion is questioned. The ego thrives in the realm of extroversion. Worldly success, sought by the ego, is achieved by a hyperactive energy, a restless emphasis on 'doing' rather than 'being', a relentless drivenness. But quite suddenly it is as if the ego's fuel supply runs out and the individual becomes exhausted by all this manic activity. As the enantiodromia is activated, a longing for something less frenetic takes hold and the pendulum swing from extroversion to introversion becomes inevitable.

The second main feature of the enantiodromia stage is the sudden withdrawal of projections. Time and again I find myself working with clients who, in their forties and fifties, are oppressed by the roles they have embedded themselves in. Invariably they have projected onto their spouse or partner someone who will give their life meaning, anticipate all their needs, heal their wounds and redeem life's deficiencies. But living with another person on a daily basis wears away these projections, as the partner can be just as wounded, needy and afraid and projects similar forlorn hopes. These are expectations we all have which we are bound to be disappointed by as our projections now in midlife appear to be misplaced. We also put a heavy projection on our role as a parent, hoping that our children will provide us with the meaning, happiness and intimacy we long for, but by midlife our sullen, challenging adolescent children quite rightly resist our projections and the expectations of parenthood are dashed against the frictions of family life. In our professional lives we seek meaning, fulfilment and achievement, but these hopeful projections – after twenty or more years – are so often overwhelmed by a sense that our profession has become a monotonous, habitual chore, a large rock, that like Sisyphus we are doomed to push up an endless gradient. As our marital, parental and professional obligations mount, we lose a sense of our true needs as we become overwhelmed by all these responsibilities. These demands, so overladen with oppressive obligation, take up all our time and energy, leaving us with no means to nourish our own deepest needs. We literally become a prisoner of these roles.

Jung experienced this intensely. The role of Freud's crown prince, the role of a brilliant psychiatrist lecturing across Europe and America, the role of the charismatic professor drawing large audiences to his regular appearances at Zürich University, the role of inspirational writer who was expected to produce books and papers for an expectant readership, the role of loving father of five, the role of dutiful and faithful husband – all these commitments crushed Jung's natural free spiritedness in a vice of overwhelming obligation. No wonder, in an act of wanton sabotage, he jettisoned all these roles and projections and disappeared from public view into a profound introversion. It was as if Jung in the previous decade had forgotten about the needs of his number two personality. His number one personality had dominated the years from 1903 to 1913, yet his number two personality was far too powerful, far too deep-rooted in Jung's psyche, to be permanently ignored. It now made a vigorous comeback.

The third stage of the midlife transition is the period of liminality, a state of being between two phases, a period of waiting, of inaction, of suspension. Jung was interested in the work of the Chinese Taoist philosopher Lao Tzu and would often recommend to his clients, when faced with an intractable problem, Lao Tzu's instruction: 'Wu Wei', which means 'let things happen'. Lao Tzu described this 'Wu Wei' as 'action through non-action', a state of stillness, a mode of waiting. In psychoanalytic terms, as the old ego-orientated life is shed, the vacuum left will not be immediately filled, all of life's pathways to the future will appear blocked and the future seems unimaginable in every conceivable direction. For Jung this 'Wu Wei' – this fallow period of liminality – is essential in order that the Self can emerge. As Jung discovered, dreams, fantasies and synchronistic events would fill the empty space, yet so many of these inner images will cause fear and anxiety. If the state of liminality is truly entered into, it means we have shed much of our ego-orientation, but with this shedding we tend to leave behind many of our carefully constructed defences of denial, projection, repression and deflection. As our defences are lowered, Winnicott's primitive agonies will re-emerge with alarming effect and we are then faced with acute anxiety, insomnia, panic attacks, eating disorders and other neurotic symptoms.

As we have seen, these primitive agonies have lain dormant in the unconscious, occasionally sending a reverberating twinge of anxiety up through the geological strata of the psyche, until they erupt during the midlife crisis. As Winnicott writes in 'Fear of Breakdown':

> The breakdown has already happened near the beginning of the individual's life. The patient needs to 'remember' this but it is not possible to remember something that has not yet happened. The only way 'to remember' is for the patient to experience this past ordeal for the first time in the present.

It is while in this phase of liminality that Winnicott suggests the primitive agonies resurface and now have to be confronted. In other words, it is as if we have to return to these earliest traumas without recourse to the defenses we deployed in our initial months and years of life. Hopefully during this most challenging of times we will be reassured and supported by a psychotherapist or empathetic friend, a reassuring, comforting presence that Toni Wolff offered Jung.

The other key feature of the period of liminality is our deeply unsettling encounter with the stark reality of our own mortality. Heinz Kohut, the American psychoanalytic theorist, writes that one of the great psychological challenges of midlife is the 'acceptance of life's transience'. As the span of years we have left narrows, death comes into sharp focus and this encounter with our mortality joins the re-emergence of our primitive agonies. This preoccupation with death presents itself in a number of ways and can create a tremendous anxiety about the health and welfare of loved ones. It can provoke spasms of chronic hypochondria, when every small ache and pain becomes a symptom of terminal cancer, and thoughts of suicide are common. We have now reached Winnicott's 'the bottom of the trough'. With death as life's final eventuality, the meaning and purpose of life appears as a pointless absurdity. Jung describes this preoccupation of death as a symptom of egocide, as it is the ego that is terrified of death, but the emerging Self, free of the ego's narcissism and grandiosity, accepts death as a natural, organic conclusion to life.

Having confronted our primitive agonies and our acute fear of death we can leave the phase of liminality and move into what we might call the 'midlife developmental phase'. During this phase two particular aspects of our true nature begin to appear in what we might describe as the 'return of the anima mother' (the anima being that element of our nature which longs for connection, relationship and intimacy) and 'the emergence of the shadow'.

In many schools of psychotherapy, and particularly in the Jungian model, there is a premise that the psyche is constantly searching for strategies to heal old wounds in order to facilitate psychological development. In pursuit of this, the psyche looks for opportunities for a kind of developmental catch-up, whereby each individual is given the chance to make up for deficiencies and inadequacies in the parenting that they experienced. If, as is often the case, the Winnicottian 'primary maternal preoccupation' has been inadequate, the individual, during the midlife transition, will often search for someone who can belatedly provide the intimacy, empathy and warmth which was absent in their prior relationship with their mother. This is the phase of 'the return of the anima mother', when someone emerges who provides an intoxicating mix of parent, lover and companion in an experience of emotional intimacy which was lacking in childhood. Because of the emotional deficiencies of his own mother and with Emma's attention taken up with raising their five children, Jung found that his intense need for this kind of belated maternal attention was provided by Toni Wolff. Her love and devotion fed an anima-longing that had remained in his unconscious since childhood and which suddenly appeared with an impulsive urgency.

A male companion, lover, husband or psychotherapist is quite capable of providing this maternal role. It can be filled by a spouse, but often the partner won't have the time and energy required because of other obligations, so the role is filled by someone outside the marriage, as Jung experienced. The dangers of the anima's seductiveness and the consequent collateral damage to marriages and families is obvious, and yet its potential for facilitating the maturation process is very considerable as the renewed anima/mother presence is full of creative developmental possibilities. To shy away or to deflect this necessary encounter will inhibit this key maturing element in the process and so the challenge is finding a path which allows this crucial developmental experience but which doesn't cause lasting damage to marriages and family life. Seeking this balance is perhaps one of the greatest challenges of the midlife programme.

The second key element of the midlife developmental phase is 'the emergence of the shadow'.

In the first half of life, one of the most significant allies of the ego is the 'persona', a concept identified and named by Jung and given a long technical definition in *Psychological Types*. It is what Jung calls 'a compromise between the individual and society as to how a man should appear to be': the psychological clothing an individual wears in public situations to avoid exposing emotional vulnerability. It is also what the sociologist and social psychologist Erving Goffman has called 'the presentation of self in everyday life'. The midlife transition will be one of the few times in life when we will completely suspend the persona, as Jung did between 1913 and 1918.

The extrovert personality inhabits the realm of the persona which is influenced by peer group pressure and tends to comply with social norms and expectations. On the other hand, the introvert personality will question this need to comply and will often suspend the persona altogether. In recent years the persona has become addicted to social media, that barrage of influences and demands to conform. It shies away from individuality, will grumble about any aspiration towards individuation, and as a result the emerging authentic self during the midlife transition will try to restrict the persona and end its impact.

One of the main functions of the persona is to constrain and silence the shadow, its ultimate foe. In contrast to the persona, the shadow is that area of our psyche, hidden out of view in our unconscious, which contains all those aspects of our nature that we have been taught to regard as shameful and unacceptable, including anger, sexuality, lust, pride, sadness, rage, jealousy, envy, hate and sloth. More often than not, any emotion or state of mind that the collective around us condemns or censures will be pushed out of sight into the shadow, which is normally ignored in the first half of life, but almost always appears during the midlife crisis. Despite their disreputable status these elements of our shadow, however much we ignore them, maintain a powerful reality within our psyches. For Jung his sexuality was perhaps the most dangerous intrusion that emerged from his shadow as it was especially offensive to the culture in which he lived.

In *Freud and Psychoanalysis* Jung writes:

> The task of psychotherapy involves a more honest encounter with ourselves and our shadow, some deepening of the journey into places we would rather not go. Our task is to live through them, not repress them or hurtfully project them on to others. To experience some healing within ourselves we are summoned to wade through our own muck and darkness. Where we won't go willingly, sooner or later we will be dragged.

My own personal experience and that of my clients has shown me that Jung is right. Invariably the experiences of the midlife transition force us to confront our shadow. Indeed, following Jung's advice it is my belief that as psychotherapists one of our most pressing tasks is to examine this dark corner of the unconscious as diligently as possible and once we have engaged with this onerous work to help our clients face this murky, unsettling terrain. If we can meet the challenges of the shadow, we will quite unexpectedly be granted a new energy source, an unanticipated gift of dynamic vitality. The encounter with our shadow and its conscious integration into the full expression of our personality is one of the most important and challenging features of the individuation process.

As these various phases of the midlife crisis are traversed, a sense of renewal and a return to a much-welcomed stability is often felt. Acute anxiety will moderate and the mood of the psyche will oscillate between a mature, more optimistic, less fearful side of our personality and the still troubled, wounded inner child that continues to crave attention and comfort. The last phase of the midlife transition could be perhaps described as the 'reparenting of the wounded child'. The other side of the client's nature, the emerging mature adult, now has the resources to soothe and reassure the inner child whenever he or she reappears. As all these various phases run their course, and we engage with their demands and pleas for attention, we will begin to sense a psychological renewal. And yet despite this the process of individuation still has much to ask of us as it reveals hidden, surprising aspects of our true nature that now emerge perhaps for the first time.

One additional consequence of the individuation process seems to emerge from the midlife crisis. Almost all my clients, as they recover, talk of something they describe as 'soulful' or 'spiritual', which becomes important to them. Perhaps we need to remind ourselves that 'psyche' is the Greek word for soul and therapy comes from the Greek word *therapeia*, which means to heal. The term psychotherapy can therefore be translated as 'healing the soul'.

Jung had a tremendous sense of the soul or the spirit. Above the front door of his house in Küsnacht he placed a quote from the Delphic Oracle: WHETHER INVITED OR NOT THE GODS WILL BE PRESENT. Since it was published in 1933, one of his most popular books has been *Modern Man in Search of a Soul.* My own clients, during their midlife upheavals, often long to develop this quality of soulfulness which they sense will offer them a feeling of inner harmony and peace of mind that helps them accept their vulnerability, their weaknesses and their own mortality. This condition of inner poise, humility, equanimity and compassion is ultimately achieved through self-knowledge.

What Jung is suggesting is that we need to set aside the supernaturalism of religion and turn towards a truly human-based spirituality. This sense that our spirituality should come from within ourselves rather than from something supernatural has, as its premise, the understanding that the prime purpose of life is to develop not only our intellectual intelligence, but also our emotional, ethical and psychological wisdom and that this depends on our capacity for self-knowledge. It is an ancient belief which first appears on the fourth century BC archway of the Temple of Apollo at Delphi in the first of the three Delphic maxims: 'Know thyself'. This is followed by Socrates' dictum, 'The unexamined life is not worth living', before it threads its way through Plato, Aristotle, Montaigne, Shakespeare, Rousseau, Coleridge, Emerson and finally to Jung who, as we have seen, places self-knowledge and self-realisation at the summit of human achievement.

– The Shadow: A Powerful but Turbulent Ally –

For some time I have been in the habit of spending a few days each year in a monastery, away from the demands of a busy life. Three years ago I intended to use this time to write a paper on Jung's concept of the shadow, which I was due to give a month later, and decided to try a new retreat, a convent where a group of nuns provide people like me with a refuge.

The week before had been especially stressful. My small company had entered into a partnership with a much larger company based in New York and a rather troubled project resulted in me having several cantankerous conversations with the American chief executive, but by the end of the week our difficulties seemed to be resolved and I set off, relishing the prospect of five days full of peace and quiet.

The convent turned out to be exactly what I was looking for and being set in a beautiful stretch of Gloucestershire countryside, I immediately felt at home. On arrival I quickly slid into a relaxed mood. I attended mass and several other offices of the day, congratulating myself on how quickly I had shed the noise and bustle of the outside world. I had a number of pleasant conversations with several of the nuns, who were warm and welcoming, and I went to bed that evening feeling certain that I had found the perfect haven. At about 2 a.m. I woke up from a deep sleep, feeling strangely alert and on edge. I turned on my phone to check the time. To my horror I saw I had three missed calls from New York and an angry email from the American CEO demanding that I travel to Brussels the following morning and sort out our problematic project that was now in a state of imminent collapse.

I felt an intense fury gather up inside me and without a moment's pause for thought rang the American CEO in New York. When he picked up, the unwanted sound of his voice pitched me into an even higher state of rage. Almost out of control I shouted at the top of my voice a great tirade of fury, lacing my language with the most abusive expletives. The furious argument seemed to go on for some minutes and ended when I said I had no intention of travelling anywhere and that I was forthwith resigning from our partnership.

As soon as I put down my phone, I realised that the window to my room was wide open and that I was surrounded by sleeping nuns and a number of visitors on silent retreat, all of whom must have been woken by my loud angry outburst. As I lay there with my anger abating, I felt filled with shame. My beautifully arranged persona, presented to the nuns with all its charm and poise, had collapsed and I had revealed myself to be an angry maniac who had woken the entire convent at this unearthly hour. The shame mounted during the next couple of hours. There was no way I could face the nuns the following morning and so I packed my bags and bolted.

Talk about heuristic research. In this tranquil place, my shadow had exploded into view and I had been shamed into this mortifying nocturnal flight. I had, in the months and years prior to this outburst, prided myself on my calm and placid response to challenging situations and had come to see anger and irritability as something that my composed demeanor had now mastered, as I had now finally shed old patterns of neurotic behaviour. Who was I kidding? I had just pushed these unseemly emotions, so unbefitting a wise psychotherapist, deeper into my shadow until this explosion of nocturnal anger had awoken a convent full of sleeping nuns.

Shadow outbreaks are often triggered by some minor incident, but their eruptive, apoplectic nature stems from something much larger and significant, some inner tension that surfaces in an explosive fashion. After my shameful departure from the convent I knew there was something that needed to be confronted and resolved and that unless this mysterious issue was addressed, I could be sure that there would be further eruptions.

As we have seen, the prime resource we use to keep the shadow in its hidden place is the persona, that mask of social conformity and collective compliance that we show to those around us. The persona is the ego's ally in the taming and imprisoning of the shadow. At some point in our middle years, when we plunge into the murky waters of our midlife crisis, we will have used up all the energy of the persona-based ego side of our personality and be searching for a different orientation as the persona loses its effectiveness. At this stage the shadow may suddenly express itself in an eruption of hate, malevolence and pernicious intent. Accompanying all this venom will be a most unsettling feeling of guilt and shame, activated by the fact we are capable of having such repellant feelings. And yet this outpouring of agitated emotion will also be a great release as it rises up from the unconscious into the light of consciousness. For years and years this psychic silt has lain in the unconscious, festering away and sending tremors of depression and anxiety up into our conscious minds. The energy and effort required to suppress all this dark effluence is tremendous, often resulting in exhaustion or somatic illness. If all this dark material can return to the light of day and somehow be integrated and accepted into our conscious personalities, then a new energy source will suddenly be available to us.

However, this process of release is full of danger. As the shadow erupts into visible presence there may be much personal and collateral damage, including the possibility of divorce, family breakup, professional collapse, emotional breakdown or illness. A successful working alliance with a therapist can facilitate the emergence of the shadow, avoiding much of this potential damage, but the shadow's release may be beyond any controlled management. As the shadow is accepted and expresses itself, the considerable energy used to suppress it can be used to fuel a more creative, more authentic vitality that is perhaps felt for the first time. The emergence of the shadow has been a key feature of the work I have done with almost all my clients.

The Emerging Shadow in the Client's Parental Relationship

During the course of the first few sessions I find that most new clients give an account of themselves and their lives using, as their main means of expression, a kind of inner critic that gives a long list of all their shortcomings. It's as though a catalogue of low self-esteem and personal failings have coagulated into a stagnating mood of depression, which is often repressed anger turned in on itself: an anger that is so shameful and unacceptable that it has to be contained and pushed deep into the psyche where it is constrained within the shadow. As I work with the client in these initial sessions, I am trying to locate the origin of this fierce inner critic that bears down on them. When did they first feel these negative feelings about themselves? Who or what gave these negative feelings their first nudge or initial momentum? Quite soon the origins of these negative feelings, with their paralysing low self-esteem, begin to emerge and most frequently the client will describe a childhood full of constant, relentless criticism delivered day-to-day by parents who were burdened with their own troubled psyches.

A familiar scenario is one in which the client's mother has experienced her marriage as an emotional disappointment and tries to compensate for this through a powerful bond with her child. But as inevitable tensions develop between mother and child, all too often the mother will become disappointed by the child's inability to give her the emotional gratification she craves, a situation which will merely extend the experience of emotional disappointment that the mother has perhaps had since her earliest days, a repeated experience that pre-programmes her response to her child. The mother's consequent despondency will all too often then ignite parental criticism towards the child that may be reinforced by a compliant husband and malleable siblings. The client's world, as far back as he or she remembers, becomes an environment of perpetual disparagement and their response to this will be to collude with all this parental disappointment and become disappointed in themselves, thereby firmly establishing their inner critic. As we explore these aspects of their parents' emotional lives, the client begins to realise that all this criticism did not begin within his or her personality but rather came from another source. As this realisation takes hold, the therapist gives permission to the client to explore feelings of rage towards the origin of this destructive criticism, the true origin being emotionally dysfunctional parents.

In the course of a few months I have often seen new clients move from compliant, respectful children to furious, enraged offspring. I remember one particular client who had unusually overbearing parents to whom she was subservient in a docile, submissive fashion, accepting all their criticism without ever questioning their acerbic judgements. Within four months of beginning therapy we had unearthed an incandescent rage directed towards her mother and father and I vividly remember her describing how she would like to plunge a carving knife deep into the ribs of both her parents. We worked with this fantasy for a number of sessions as her anger erupted out of her unconscious and the oppressive power of parental prejudice was now breached.

What we are dealing with here is the battle between the client's emerging shadow and his or her parents' 'superego', the set of cultural rules which the parents have been indoctrinated in which will then impact on their children by means of what we might call the 'family superego regime'. This family superego regime can range from being overbearing, judgmental and oppressive to flexible, tolerant and magnanimous. The more oppressive the parental superego regime, the larger the child's shadow will become. A censorious parental superego regime will be used to control the child in a fashion that suits the needs and prejudices of the parent, but will not serve the needs of the child.

The client, as we all do, needs a superego as an ethical guide and will invariably be drawn towards the therapist's superego regime, a completely new set of values that stand in opposition to those of the parents. However, the therapist has to resist imposing his or her ethical principles on the client and should be trying to tease out the client's true feelings that have been overwhelmed by the parents' superego. Invariably the only power sufficiently energetic to enable the client's real feelings to emerge is the force generated by the shadow; only the shadow can empower the client's triumph over the parents' superego. This sudden change to a client's worldview can be as frightening as it is liberating and I have seen many clients lapse back into the world of the inner critic and the parental superego regime. But once the shadow has been given the life-giving oxygen of consciousness, the anger, rage and consequent energy of the shadow will never return into the dark cave of the unconscious. Once expressed it is too powerfully present to be once more constrained.

Also, the therapist must teach the client what we might call the 'economy of the shadow', for it has such an eruptive energy that it is capable of causing havoc. The client's partner, children, professional colleagues and other family members who have nothing to do with the original parental offense may come into the firing line with disastrous consequences and part of the role of the therapist is to pace the emergence of the shadow, getting the client to act out the shadow's fury within the safety of the therapeutic sessions. Many of my clients in their forties and fifties have had ageing parents who have long since shed their hectoring, critical tendencies that had been so detrimental, but I have often felt it necessary to suggest that these clients restrain from delivering a furious tirade in the direction of elderly, frail parents in their late seventies or eighties. Yet sometimes the desire for revenge is too great and the client will act out their overwhelming need to express these furious emotions in front of ageing parents.

Generational Family Shadows and their Corrosive Impact

In *Memories, Dreams, Reflections* Jung writes:

> I feel very strongly that I am under the influence of things or questions which were left incomplete and unanswered by my parents and grandparents and more distant ancestors. It often seems as if there were an impersonal karma within a family, which is passed on from parents to children. It has always seemed to me that I had to answer questions which fate has posed to previous generations and which had not yet been answered or left unfinished.

Indeed, my own family has a tendency to repeat a lethal pattern of replicated shadow experience, selecting alcoholism and suicide as their preferred calamity. My paternal great-grandmother drank herself into an early grave. Two of her sons, my great uncles, took their own lives. In the next generation my aunt took a fatal overdose of sleeping pills and was followed by her son, my first cousin, who died of chronic alcoholism in his mid-thirties. This pattern was never mentioned, never commented upon: it was too shameful – too unacceptable to be confronted – and it remained deep in our family shadow. As Jung suggests, I have often felt that one of my key life tasks has been to confront this life-threatening family shadow and not only to avoid its lethal impact myself, but also to prevent it being passed on to yet another generation.

The existence and repercussions of generational shadows is present in almost all my work with my clients. I had one client whose great-grandfather and three of his brothers died in 1916, during the Battle of the Somme. A hundred years later we explored the impact of these deaths, which had so affected the lives of her great-grandmother, her grandmother, her mother and finally her own. All four generations of women experienced their relationships with men through the lens of this crippling expectation of loss that had established itself as an intensely harmful family shadow. They all seemed to labour under the unconscious premise that somehow all their men would abandon them through illness, early death, infidelity, desertion or just standard marital breakdown. It was as though the terrible events on the Somme had entered the family's psyche, creating a shadow that had become part of the family's emotional DNA. Following Freud's theory of repetition, the family's negative assumption fermented the same predictable pattern in my client, and her own midlife crisis emerged after a series of relationships had ended in rupture and separation, maintaining her allegiance to the family shadow.

In my practice I have seen so many instances of this tension in the psyche between the compulsion to repeat and the urge to develop, thereby shedding its continual pattern of the family shadow. As we have seen, while Freud's theory of repetition initially holds the high ground as the dominant position in all early clinical work, the opportunity provided by the midlife crisis to overcome the repeated pattern of the family shadow can initiate a new way forward, a new attitude that breaks the ingrained cycle of misfortune. To watch a client make this kind of progress makes the years of psychotherapeutic effort worth every session. One such case was my client Jane, who arrived for her first session six months pregnant with her first child at the age of thirty-seven. She was overwhelmed by a pervading anxiety that she was emotionally ill-equipped to become a mother and her pregnancy had become her midlife crisis. In our early sessions it emerged that she was the seventh and last child of a large family. At birth she had been unwelcomed and neglected by her exhausted mother who, in addition to raising six other children, was also working forty hours a week. Then, at age ten, Jane had been sexually abused by her father.

Fourteen years before she was born, her father had been involved in an infamous murder case and was very nearly charged as an accessory and the public humiliation and the trauma of the case had a devastating impact on his family and his career. As Jane later wrote: 'The murder engulfed our family and set in motion a way of being and relating that ended decades later with a fractured, traumatised family, siblings living in isolation from each other and parents driven to suicide.' Jane's family had lived in an all-embracing mood of secrecy, shame and alienation, dominated by repeated experiences of sexual abuse, abortion, bleak emotional austerity at home, business collapse and finally – at different times – the suicide of both parents. No wonder Jane felt concerned about her capacity to become a loving mother.

During our sessions we discovered that Jane's psychological state of being was articulated most effectively by Winnicott's famous maxim: 'It is a joy to be hidden, but a disaster not to be found.' Jane's life began with what we might call the 'trivialisation' of her birth. None of Jane's siblings even knew that their mother had been pregnant with Jane and one of her brothers described arriving home from school one day to find, quite unexpectedly, that he had another sister.

It was as if the pregnancy had been yet another shameful secret and for Jane it was as if her birth was an unwanted distraction to the daily routine. So, no time for an exchange of oxytocin, no time for some crucial bonding, no time for some breastfeeding, no time for some Winnicottian 'primary maternal pre-occupation', as Jane was farmed out to an elder sister.

This neglect stalked Jane throughout her childhood and early adult life and she found comfort in being hidden, taking solace in her secluded, withdrawn emotional state. In this family, where the need for closeness and connection and the expression of intimacy were locked away in the family shadow, she neither sought nor expected attention and after the abuse she suffered at the hands of her father, she lost her capacity to trust others. She took these feelings into her early adult relationships, experiencing a series of casual liaisons, void of any true emotional intimacy. As she later wrote: 'It was a paradox – I wanted to be seen but was too frightened to allow it.' Having never felt she was loved, her own ability to love was suppressed, existing as a latent dormant potential waiting to be discovered deep in her unconscious. Inevitably she found exactly the right husband, who fitted into this pattern. Like Jane, her partner had come from a family with similar issues around shame and sexual abuse, and who was later diagnosed with Asperger's syndrome. His condition meant he paid Jane very little emotional attention and was seemingly disinterested in her state of mind. Jane could safely remain hidden, loyal to the family shadow to which she was so committed.

In our work together she experienced a level of attention, concern, care and empathy, within the safety of the therapeutic frame, that seemed to have an immediate effect and when her daughter was born she showed all the loving maternal qualities that her own mother had lacked. After this most important of happy developments, our work together began to gather pace and with great courage, Jane strove to overcome the legacy of her family's wretchedness and the way it had sculpted her early adult life.

Jane was single-mindedly committed to her therapy. Along the way she trained and qualified to become a psychotherapist and with remarkable speed she built up a thriving practice. Her marriage inevitably faltered, after which she began a new relationship unencumbered by all her fears and anxieties. Her desire to remain hidden faded as her feelings of trust in her new partner allowed her to experience a liberating expression of her true nature and with this experience it was as if the power of her family shadow was finally broken. Now she could enter the realm of her true self.

The Family as a Container of the Shadow

A good many families will unconsciously place an embargo on overt expressions of the shadow. Anger will perhaps be prohibited or sex will be a subject that is never mentioned and often there will be family secrets that are strictly concealed. As with Jane, expressions of love or physical contact will be discouraged and criticism of the parents may be forbidden. I have known families where there is an unspoken rule that only one of the parents is permitted to express anger, while the rest of the family often meekly complies with this strict rule.

In contrast, families that have disputes, disagreements, arguments and rows and permit the expression of anger or regressions back to childhood feelings of hatred, may provide a level of shadow expression that is good for the mental health of all the family members. Personally, I believe one of the great gifts a parent can give a child is permission to express negative emotion, particularly against parental authority. For instance, the fifteen-year-old son of one of my clients told her to 'fuck off' before charging off to his room where he remained for the next three hours. Later that evening he apologised and she forgave him and expressions of love and affection ensued. This reminded me of an incident when my own fourteen-year-old daughter called me a 'fucking wanker' and I replied with an even more offensive remark. It took twenty-four hours before apologies, forgiveness and expressions of love were exchanged, which was a clear example of Melanie Klein's theory of the 'depressive position', which requires us to bear great psychological discomfort when we feel hatred towards someone that we love. This ambivalence is very unsettling, but acceptance of it is, for Klein, a sign of psychological health and developing maturity. Families where negative emotions and eruptions from the shadow are permitted provide a place where the shadow can be expressed and contained.

It is much harder for friendships or professional relationships to survive bursts of anger and if you tell a friend or professional colleague angrily to fuck off, there is every chance the relationship won't survive the insult. A relationship will only survive this kind of hostility where there is a very strong attachment and of course it is the family unit that tends to provide us with our most durable, robust attachments that can tolerate moments of anger.

The Shadow in Romantic Relationships

In my practice I'm often struck by the number of clients who arrive with their central presenting issue being their inability to sustain relationships but who then end up in a new relationship within six to eight months. The new relationship, with a good flow of oxytocin and dopamine, is quickly idealised and immediately becomes a wonderous miracle, transforming the client's life.

I, of course, don't puncture this moment of pleasure and joy, but I know that the quality and durability of the relationship shouldn't be gauged by the initial measure of happiness but by the manner in which the couple cope when the shadow of the relationship is eventually encountered. The more intense the idealisation, the deeper the shadow. The projected idealisation – casting the loved one as perfect, as flawless, as exquisite – will inevitably subside as the flow of oxytocin becomes less reliable and the ecstasy of falling in love is replaced by a suspicion that the loved one is all too human and has an increasing number of highly unattractive faults. What is being faced is the loved one's inevitable shadow.

How a couple deal with this stark realisation is the deciding factor as to whether the relationship succeeds or fails. In many ways a long relationship or marriage is primarily an exchange of shadow material and the success of the marriage will depend on whether this shadow material can be tolerated and then compassionately accepted. The new relationship will bring a collision of two shadows that may survive a long co-existence or may simply be too uncomfortable to sustain. During this period of shadow recognition, as the romantic idealisation falters, the therapy can play a vital role in whether or not the relationship can survive this shadow confrontation, and eventually prosper.

The notion that the greater the couple idealise their relationship, the larger the shadow, was highlighted in a 2014 study by Andrew Francis-Tan and

Hugo M. Mialon who found evidence that the greater the cost of a couples' wedding, the more likely was the chance that the couple would get divorced. Couples who spent $1,000 or less on their wedding were 53% less likely to get divorced than the national average, while those spending $20,000 or more on their wedding were 46% more likely to get divorced. More often than not, a wedding represents the persona of the marriage. As a large persona will always mask a large shadow, it seems highly likely that a large wedding will mask hidden fault lines in the marriage that will increase the chance of marital breakdown.

Bringing in the Harvest: The Discovery of the Positive Shadow

In the same way that nuclear power can, if contained within the safety of a nuclear reactor, provide electricity for an entire city, but if left uncontained will destroy it, the burst of energy released when shadow emotions are freed can be re-energising or destructive, depending on how carefully they are managed. Aspects of our unconscious that might be enlivened can include repressed talents and potentialities that we have neglected or denied due to some fear or repression or because of the disparaging presence of our inner critic. These repressed talents are aspects of what we can call the 'positive shadow.'

I have had numerous clients who, as they process the full impact of a midlife transition, discover a neglected talent or passion that now makes a belated appearance. Such examples include: the art dealer who came to realise that supporting other people's creativity was a compensation for his own repressed artistic talent and who, in his fifties, gave up his business and became a successful painter; the psychologist who had always wanted to write a novel but whose highly critical father created within her an inner critic that prevented her from ever writing until therapy released her literary talent; the psychiatric nurse, obsessed by his passion for classical music, who finally found enough courage to leave his profession and became the leading programme-note writer in the UK; the psychotherapist who had repressed his poetic gift in his early twenties but rediscovered his considerable talent and became a successful poet; the music critic and broadcaster who, in his late fifties, transformed himself into a talented composer.

The Early Appearance of Jung's Shadow

When I first travelled to Basel and visited its cathedral, I was struck by the close proximity of the Basel Gymnasium, where Jung was a pupil. While attending the school the eleven-year-old boy began to have a most unsettling daydream. Despite his attempts, he simply couldn't suppress an image of God sitting on a great throne-like lavatory and producing a vast turd that fell directly onto Basel Cathedral, causing the roof to collapse.

The violence and profanity of this image upset the young Jung, but then he began to wonder whether it was God's will which made him continue to have these thoughts. This released the young boy from his fear of damnation and he began to feel that perhaps he was somehow special, given to these strange fantasies and reveries that none of his other contemporaries had. This seems to be Jung's first encounter with his shadow, this experience of being tormented by this shameful image which was totally abhorrent and unacceptable in the household of a Christian priest. But he then accepts the sacrilegious thought as God's will, prefiguring his later view that the shadow is an important aspect of our true nature and should always be accepted as a life-enhancing asset rather than a frightening liability.

– Two American Presidents –

Following the Declaration of Independence in 1776 and the subsequent American Revolutionary War, the United States – in its first 170 years – was to encounter three crises that threatened its very existence. The Civil War, the Great Depression and the Second World War were national ordeals of daunting calamity through which the nation prevailed and moved forward towards its eventual position as a world superpower. Two individuals provided exceptional leadership through these historic challenges. Abraham Lincoln piloted the nation through the catastrophe of the Civil War and then seventy years later Franklin Delano Roosevelt navigated the country through the twin convulsions of the Great Depression, followed immediately by the Second World War.

Yet in both cases the two greatest presidents in American history entered their middle years displaying no signs that they might develop personal qualities that would save their nation from terminal decline and collapse. These emerged after they both lived through their own severe midlife crises.

In 1849, Abraham Lincoln returned home to Springfield, Illinois, a defeated, single-term congressman whose two years in Washington had made virtually no impact on the national political stage. It seemed to Lincoln to be yet another failure to add to the long list of other failures that characterised his career. As Lincoln's future secretary John Nicolay wrote in his memoir of the President:

> He went into the Black Hawk War as a Captain and came out a Private. He rode to the hostile frontier on horseback and trudged home on foot. His store 'winked out'. His surveyor's compass and chain, with which he was earning a scanty living, was sold for debt. He was defeated in his first campaign for the legislature, defeated in his first attempt to be nominated for Congress, four times he was defeated as a candidate for Presidential Elector.

His legal partner William Herndon later commented how Lincoln found these setbacks very painful and that he was 'keenly sensitive to his failures' and any mention of them made him 'miserable'. Many of his friends commented that Lincoln, after returning home, fell into a deep depression with 'a sadness so profound that the depths of it cannot be estimated by normal minds.' His friend Henry Whitney said that 'he was ingrained with a mysterious and profound melancholy' while Josef Fifer recalled that 'Lincoln's face was about the saddest I have ever looked upon. The melancholy seemed to roll from his shoulders and drip from the ends of his fingers.'

As a psychotherapist it seems to me that Lincoln's intense depressive tendency was caused by the repeated experience of loss which he experienced as a child. His brother died when Lincoln was three but much more devastating was his mother's death when he was nine, followed a few years later by the death of his sister. We know from numerous psychological studies that such deaths, and particularly that of the mother, represent a catastrophic loss of such proportions that it lays down in the psyche a pronounced tendency towards severe depression, especially if the surviving parent is unable to offer consoling love, affection and support. Lincoln's father reacted to his wife's death by undertaking a prolonged expedition down the Ohio River to sell pork and he left his two children with a young cousin, Sophie Hawks, who was unable to provide the emotional and physical care that the young Lincoln so desperately needed. One neighbour reported that the Lincoln children 'became almost nude for the want of clothes and their stomachs became leathery for the want of food.' Lincoln came to loathe his father, was estranged from him throughout his adult life and neither visited him when he was dying nor attended his funeral.

The sense of failure and dejection that Lincoln experienced on his return from Washington was made much worse in 1850 when his three-year-old son Eddie died. This dreadful further loss brought back all the misery of his childhood. I have often noticed amongst my clients how a death in adult life very often reconfigures a bereavement that was experienced when they were young children, returning them to the grief they had blocked in their early years. This now happened to Lincoln, with Eddie's death returning him to the trauma and primitive agonies he suffered when his mother died. Given the intensity and depth of Lincoln's depression by the end of 1850, how was it possible that this broken, defeated, failure of a man could, in the course of ten years, transform himself into the greatest of all American presidents? This remarkable change could not have happened without the influence and impact of a young woman who appeared in Lincoln's life in the winter of 1819, when he was ten.

When Sarah Bush Johnson became Lincoln's stepmother she immediately developed a close bond with her young stepson and within a few weeks he was calling her 'mother'. Although Sarah already had three children of her own, she became devoted to Abraham. She could tell that he disliked the hard labour involved in farm life and seemed to much prefer 'reading, scribbling and writing' and it wasn't long before Sarah was able to anticipate his taste in books and provide him with exactly the right material to develop his precocious mental gifts. After he had finished the King James Bible and Shakespeare's plays, Sarah introduced him to *Aesop's Fables*, Bunyan's *The Pilgrim's Progress*, Defoe's *Robinson Crusoe* and the autobiography of Benjamin Franklin.

Lincoln was never formally educated, having only had eleven months of schooling, yet spurred on by Sarah he became perhaps the most famous autodidact in history and future friends and colleagues were amazed by the extent of his encyclopaedic knowledge, his learning and erudition. With Sarah's love and encouragement, the young Lincoln demonstrated an unusual sense of self-discipline and an exceptional conscientiousness, which developed into a strong ambition tempered by a deep sense of morality and personal integrity. It was these considerable personal qualities that had been honed during his adolescence, which now came to his rescue in 1851 while becalmed in his midlife crisis.

After eighteen months of depression and grief he began to emerge from his melancholic state of mind. Believing that his mental faculties, like muscles, could be strengthened by rigorous exercise, he developed a programme of intense reading and study and entered a period of self-examination, asking himself pivotal questions. What did he really want out of life? Was the structure of his life satisfactory? What kind of legacy did he wish to leave? Had he paid too much attention to the demands of those around him and conformed to their pressure? What did he care about most? What were his basic beliefs? Had he suppressed parts of his personality that he now needed to develop? How should he deal with the uglier, less attractive aspects of his personality?

Concerned about how little he had achieved, Lincoln told his friend Joshua Speed that he had done nothing to make any human being remember him and that his name was unconnected to the events transpiring in his day and generation. With these preoccupations about his legacy in mind, he became obsessed with his mortality, anxious that death didn't intervene before he had left his mark on the world.

All this introspection produced in Lincoln an awareness of his own uniqueness that bred in him a powerful self-confidence which inspired confidence in others. This process also helped erode his narcissism and egotism. He no longer took things personally, he accepted his own and other people's shortcomings with a sympathetic, almost amused equanimity. As the 1850s progressed, his friends and colleagues noticed a dramatic transformation in him. Joshua Speed reports a new quality that he exuded as he recovered from his depression:

> If I was asked what it was that threw such charm around him, I would say it was his perfect naturalness. He could act no part but his own. He copied no one in either manner or style. He had no affectation in anything . . . His whole aim in life was to be true to himself and being true to himself he could be false to no one.

Lincoln's friend and political ally Joseph Gillespie, observing him in 1858, talked of his 'psychic radiance', which produced 'a magnetic influence that was inexplicable, which brought him and the masses into a mysterious correspondence with each other.'

This quiet charisma, this common touch, impressed the railroad conductor Gilbert Finch who wrote of his encounter with Lincoln:

> He put on no airs. He did not hold himself distant from any man. Yet there was something about him which we plain people couldn't explain that made us stand in awe of him. But you could get near him in a sort of neighbourly way, as though you had always known him, but there was something tremendous between you and him all the time.

Here Finch captures that empathy – that absence of self-regard – that an individuating soul, like Lincoln, radiates. Any last vestige of narcissism was eliminated, leaving a powerful, radiant empathy that inspired others, which is very much in line with the psychoanalytic theorist Heinz Kohut who states that the prime feature of wisdom and maturity is the move from narcissism to empathy. Leo Tolstoy wrote of Lincoln: 'He was what Beethoven was in music, Dante in poetry, Raphael in painting and Christ in the philosophy of life.'

With this remark of Tolstoy's in mind, it is ironic to reflect that Lincoln was assassinated on Good Friday, 1865. The grieving sermons on that Easter Sunday started the process of turning the dead president into an American Christ. As the funeral cortege returned by train to his hometown of Springfield, Illinois, the journey from Washington was interrupted time and time again as each city through which they passed demanded time with the dead president. The process of deification had begun.

When we see a man like Lincoln falter and collapse into a state of inner turmoil and misery that we also fear, it is reassuring to observe his miraculous recovery and although we know we cannot emulate such a man, his example can help us through such moments. As we have seen, prior to his midlife crisis, Lincoln had achieved almost nothing and castigated himself for his repeated failures in life. In contrast, Roosevelt had enjoyed a decade of continual success and his high self-regard, at the age of thirty-eight, veered towards a condition of hubris before he encountered his midlife ordeal.

Much like Jung, FDR had become a young man of enormous promise and encouraged by his wife and mother, he was keeping up with the demanding schedule set by his cousin Theodore, the twenty-sixth president of the United States, whom he was determined to emulate. He, like Theodore, had become a member of the New York State Senate in his twenties. Then in 1933, at the age of thirty-one, he had become a junior member of President Wilson's Cabinet when he was appointed Assistant Secretary of the Navy eight years earlier than his cousin. He now had his eyes set on becoming State Governor of New York, just as Theodore had, which would provide him with the right launch pad for the Presidency.

As with Jung, there had been some domestic upsets on the way. In 1914 FDR had fallen in love with his wife Eleanor's secretary, Lucy Mercer, and he had her transferred to the Navy department where she became his assistant. The affair that followed continued until late in 1918 when FDR, returning from a visit to the European battlefields, came down with Spanish flu and was confined to his bed, leaving Eleanor to unpack his luggage. There, amongst his belongings, she discovered a box full of letters in a hand she immediately recognised. Distraught, but as always magnanimous, she offered FDR a divorce so he might marry Lucy Mercer, but his mother intervened and warned her son that if he considered such a move she would disinherit him and his political ambitions would be at an end. FDR, a father of five, was left with no alternative and promised his mother and wife that he would never see Lucy again.

FDR and Eleanor's marriage survived, but it seems the tension and strain between them at this difficult time probably contributed to the disaster that befell FDR two years later. The strained relations not only with his wife, but now also with his mother, had FDR working even harder than ever and in the 1920 election he ran as the Democratic candidate for the vice presidency, even though he was only thirty-eight. Although the Democrats lost, FDR's charm and charisma had made a great impression and he now looked forward to running for president in 1924. But for now Roosevelt returned to Hyde Park, the family's mansion on the Hudson, utterly exhausted.

Each summer the Roosevelt family would spend six to eight weeks holidaying on Campobello Island in New Brunswick. During the summer of 1921, FDR was out sailing when the family spotted a small forest fire on one of the lesser islands that surrounded Campobello. With great enthusiasm they set to their task and using pine boughs succeeded in extinguishing the blaze and returned home delighted by the day's adventure. But that evening FDR began feeling unwell. A debilitating exhaustion came over him, he began to shiver uncontrollably and an ache deep in his back intensified. He went up to his room and closed the door. Climbing that flight of stairs on 10 August 1921 was the last time he walked unaided and the next day he was diagnosed with poliomyelitis, or polio.

Three weeks later, when the family returned to New York, FDR was in a dreadful state both physically and psychologically. To lie in bed, hour after hour, day after day, was torture for this energetic, restless, exuberant man of thirty-nine; for someone of such elegance and poise to have to call for help every time he wanted to urinate or defecate was an unbearable humiliation. Having lived such a charmed and successful life, suddenly to be paralysed was a torment almost beyond bearing. No one knew what kind of life might be possible for him but one thing seemed certain: his political career was over. His mother decided that the best thing for her Franklin was to come home to Hyde Park and live the life his invalid father had lived. She would take care of him and he could pursue his hobbies, such as stamp collecting. But Eleanor knew this would destroy FDR. During this desperate period she became deeply attached to Louis Howe, FDR's political manager, who felt sure that FDR should not be treated as an invalid and remained committed to FDR's career in politics. 'I believe one day he will become president,' he told Eleanor.

With Eleanor and Louis Howe's encouragement, FDR refused to accept his doctors' verdict that he would never walk again and he entered a seven-year limbo, or state of liminality, where his sole focus was to re-learn to walk. He told his mother that rules that applied to other people did not apply to him and, as if to prove her wrong, he tried everything – daily massage, saltwater baths, ultraviolet light, electric currents, special electric belts and pulley arrangements. His upper body grew immensely strong, but his legs remained withered and emaciated. His daughter Anna remembered him saying: 'I must get down to the end of the driveway, all the way down the driveway.' Each day, with the aid of crutches, he made the attempt, but he never managed to reach the gates onto the Post Road, at the end of the Hyde Park Drive.

Polio was entirely different from all the other

challenges that FDR had faced. His natural charm, his cunning, his persuasive tongue were all useless to him now and this dreadful paralysis demanded qualities from him that had always been in short supply in his previous cosseted and privileged life, notably patience, considerable powers of self-discipline and acceptance of his own physical disability. By 1924 there seemed to be no improvement. He was finding it harder and harder to keep a brave face in front of his family as he was hit by waves of depression. As well as his faltering mood, he was finding the fraught hostility between his wife and mother, as they battled for control over him, increasingly unbearable. He had to escape.

On 2 February 1924 he was helped onto the overnight sleeper to Jacksonville, Florida, and two days later he boarded an old houseboat called the *Larocco*, which was to become his home for much of the next two years. The boat was skippered by Robert Morris, whose wife did all the cooking, but chief among FDR's small entourage was his devoted secretary, Marguerite 'Missy' LeHand, who had been in love with him ever since she had joined his staff four years earlier at the age of twenty. Now safely away from his family, FDR was able to shelve all that stoic self-discipline and fall into a deep depression. Those who accompanied him during all these listless days tell of a great deal of anger, of grieving, of frustration and a great deal of alcohol. Yet the *Larocco*, on the shores of Florida, became a haven where he could mourn the loss of the body that had once been his. As Missy LeHand said: 'There were days when he couldn't pull himself out of his depression until noon, when finally he felt he could face the rest of us.' This was FDR's phase of liminality, when he had to grapple with the full ferocity of his primitive agonies.

For much that year FDR battled with his demons, but in the autumn he made his first visit to Warm Springs, Georgia, a small impoverished farm community ten miles from the nearest paved road. On the edge of the village stood the dilapidated Meriwether Inn, a rambling Victorian building which had been popular with Atlanta society fifty years before. Its past success had been based upon a gushing hot spring which rose out of the local hills and FDR had read how a previous generation of polio sufferers had come to Warm Springs to seek comfort and recuperation in its healing waters, which were known locally as the 'miracle waters.'

FDR immediately fell in love with Warm Springs and decided to restore the original charm of this now forgotten spa. He wrote to his mother: 'I feel a great cure for infantile paralysis and kindred disease could be well established here' and persuaded her to lend him enough money to completely refurbish and restore the hotel, spa and numerous estate cottages that had once housed the guests. Within two years he had established the first 'Treatment and Rehabilitation Centre' specifically designed for polio in the US and when it opened, he devoted himself to its success and became both the clinic's director and its first patient. He wrote to a friend: 'You would howl with glee if you could see the clinic in operation with the patients doing their exercises under my leadership.' Although he was a patient, the other patients and staff called him Dr Roosevelt, but after a time this formality was dropped and he was simply called 'Rosie'. Each day, after exercise and treatment, the children and adult patients would go to the play pool where all sorts of fun and games were led by clinic director Rosie.

If you look at the photographs and films of FDR on the *Larocco* and then compare them to those taken at Warm Springs, it is as if you are looking at two quite different men – one miserable, defeated, close to despair and then the other jovial, relaxed and completely at ease with himself and his surroundings. The sense of transformation is phenomenal. FDR, like Jung at Bollingen, had found his spiritual home at Warm Springs. In later life, whenever exhausted by high political office, he would return to recover and replenish amongst the people of the local community and the staff and patients at the clinic. As one local resident said: 'Everyone loved him. It went beyond liking, they loved him.'

In the summer of 1928, reinvigorated in soul, mind and body, he decided to re-enter politics. Eleanor began telling everyone that he would be taking part in the Democratic National Convention in Houston that year, where somehow he would have to walk to the podium and nominate the governor of New York, Al Smith, as the Democratic candidate for the forthcoming presidential election. When the day arrived fifteen people watched as FDR made his way to the podium, using a stout cane in one hand and with the other hand precariously balanced on his son Elliott's arm. He appeared to be walking, yet he knew that if he fell his political career would be over. A reporter at the convention wrote: 'Here on the stage is Franklin Roosevelt, a figure tall and proud, even in suffering, pale with years of struggle against paralysis. A man softened, cleansed, illuminated by pain. The sight of this man lifts us up.'

FDR had survived his midlife crisis. He was now transformed. Six months later he became state governor of New York and four years after that he became one of the greatest of all American presidents, strengthened and tempered by the sheer extremity of what he had been through. Sculpted by his own crisis he was now imbued with a compassion, humanity and empathy which allowed him to lead his own country – and later the combined Western democracies – through the ordeals of the Great Depression and the Second World War.

– Marie Curie: The Sister of Prometheus –

As the nineteenth century progressed into its final decade there was a strong feeling, within scientific circles, that most of nature's building blocks had been discovered, measured and catalogued. This sense of smug satisfaction was most strikingly represented by the periodic table which gave the firm impression that a virtually complete tabulation of the elements had been reached and that this inventory would now remain fixed and unchanging forever.

This feeling of certainty was to be shattered when a mere graduate student, an obscure young woman from Poland, made a series of discoveries that would transform humanity's understanding of matter. So significant were her findings that in 1904 she became the first woman ever to win a Nobel Prize. Since 1901, 391 men have been awarded Nobel Prizes for physics and chemistry, while only ten women have achieved this distinction. Then in 1911 she won a second Nobel Prize, becoming the only person to win two such prizes for physics and chemistry. Her name was Maria Skłodowska, but she became known throughout the world as Marie Curie.

Of all the midlife crises that I have both studied

and witnessed, the one suffered by Curie when she was thirty-nine was the most prolonged, intense and heart-rending that this book will consider. The impact of its hardship and misery didn't let up for eight years and at first sight appears to have had no positive aspects or benefits. In addition, it's highly likely that the menopause overlapped with Curie's midlife crisis, which may have exacerbated the adversity she experienced. Because the menopause is a physiological transition that isn't directly related to the psychological conversion of midlife that this book examines, it's outside the scope of our discussion, but it would be remiss of me not to mention that studies show that when the menopause does coincide with a midlife crisis, a woman's experience is often profoundly more difficult than it would have otherwise been.

It is ironic that this Polish woman succeeded in her scientific work in such a heroic fashion at a time when Poland didn't exist as a country, when almost all countries refused women entry into university and when Western democracies didn't even allow women to vote in democratic elections. One of the only countries where women could attend university in the 1890s was France and as a result, many women travelled to Paris from all over Europe to begin their higher education. Leaving her own country that had been entirely absorbed by Germany, Russia and Austria, Maria Skłodowska arrived at the Sorbonne in 1891, aged twenty-four. During her early student days she met a rather dreamy, absent-minded graduate student. It was to become a devoted relationship that turned into a scientific partnership of historic significance. In 1895 Pierre Curie and Maria Skłodowska were married and by 1897 Curie began her work for her own doctorate, a high ambition given the fact that no woman in France had ever received a doctorate in physics or chemistry.

At that time physicists throughout Europe had become fascinated by the phenomenon of X-rays. Although Curie was also interested in X-rays, her doctoral thesis focussed on the rays given off by the element uranium, which had been observed by Henri Becquerel before he had discontinued his work, leaving this field of investigation open for Curie. She, however, had a major resource not available to Becquerel, for Pierre had invented a device that could accurately measure tiny amounts of electricity. With Pierre's measuring apparatus Curie discovered that uranium had the unique ability to charge the air around it with electricity. This leaking of electricity out of particles of uranium into the surrounding air was a completely new observation and Curie decided, using Pierre's apparatus, to see if any other elements had this property. For some time her results were predictably dull – it seemed that no other element had this electrical feature – but then in February 1898, in the course of a single week, Curie made two startling discoveries. Firstly, she found that the element thorium could also electrify the adjacent air, which led her to consider the possibility that this was not a property of a specific element, such as uranium, but rather a property of matter itself. She named this phenomena 'radioactivity'. Secondly, she discovered that pitchblende – the name given to uranium ore – turned out to be four times more radioactive than uranium. Initially she was extremely sceptical and repeated the experiment time and time again before testing Pierre's measuring equipment to see if it was faulty, but no errors could be found. Curie and Pierre began to consider the possibility that the pitchblende might contain some new, as yet undiscovered element. This seemed to be the only possible explanation for the pitchblende being four times more radioactive than uranium.

Curie then produced a paper to describe her findings, which many scientists have since called the most important paper in the history of chemistry. Yet this little-known paper by a foreign student – and even more outlandish, by a woman – was completely ignored. Nevertheless, Curie was certain of the immense significance of her findings as was Pierre, who dropped his work on crystals and joined her in the experimental work that would be required to prove her theory. By July 1898 Curie and Pierre realised that they had discovered two new elements which they named polonium (after Poland) and radium. This second mystery element appeared to be 1,000 times more radioactive than uranium, but once again the announcement of the arrival of radium and polonium was greeted by almost complete indifference. Curie and Pierre had years of back-breaking work in front of them to establish the certain existence of their two new elements. They began by renting a dilapidated old shed and imported large quantities of pitchblende from a mine in Bohemia. Using great cauldrons, they broke down and heated the pitchblende and by mid-1902, after four years' toil, they had succeeded in isolating a 10th of a gram of radium from ten tons of pitchblende. With this minute amount of the new element, Curie was able to measure the atomic weight of radium at 225.9, very close to its future official weight of 226. She also positioned radium accurately in the periodic table.

Curie and Pierre were vindicated. Radium and polonium now officially existed, as did the phenomenon that Curie called radioactivity, because of the superhuman obstinacy of this female graduate student. In 1903 she became the first female scientist to receive a doctorate in France and a year later she became the first woman to be awarded a Nobel Prize. Initially the Nobel Foundation had given the prize to Pierre alone, but he said he would refuse it if Curie was not also included in the honour. Over the next two years, Pierre's career began to flourish. He became a professor at the Sorbonne and was then elected a fellow of the prestigious French Academy of Science, two honours denied to Curie on account of her sex. On 5 July 1905 a reporter from the newspaper *La Patrie* went to the Curies' house to interview Pierre about his election to the academy. The reporter wrote:

> Unfortunately I arrived to find that the eminent scientist had gone off to visit his new colleagues at the academy. In his absence I was received by his admirable collaborator, Mme Skłodowska Curie, who rejoiced at her husband's success. When I asked whether she might receive a similar honour she dismissed the question by saying, 'Oh me, no, I'm only a woman.'

In May 1906, the Sorbonne would relent when Curie received the unique distinction of becoming the first woman to become a university professor in France. Yet this historic moment occurred as a result of a horrific accident. On 16 April Pierre had absent-mindedly crossed the Rue Dauphine in heavy rain when he was run over by a large, heavy wagon, the rear wheel of which crushed his skull, killing him instantly. So began years of intense, emotional pain for Curie, a kind of calvary of anguish and desolation from which she never completely recovered.

One is left to wonder whether the extremity of her grief had its origins in the traumatic losses of her childhood. Her sister Zosia had died when Curie was six, but worse still was the death of her beloved mother when she was ten. As we have seen, Winnicott's concept of primitive agonies suggests that a child, when confronted by the death of a mother, will repress the traumatic pain deep into the unconscious. When a similar bereavement is experienced in adult life, the present loss will be amplified by the return of these childhood primitive agonies and will activate a profound depression, similar to that suffered by Curie. This conjunction of the two traumas, one suffered by the child, the other by the adult, becomes almost unbearable.

However, in the spring of 1910 Curie appeared to make a most unexpected recovery. It was as if whatever had intervened had cauterised the vein running up from the unconscious, preventing the primitive agonies of her childhood losses from increasing the wound of her bereavement. With these agonies staunched, the pain of Pierre's loss could be soothed and eased by the new redemptive event that Curie was now clearly experiencing, resulting in a transformation in her brain chemistry. Suddenly her amygdala stopped drenching her brain with adrenaline and cortisol and her dopamine and serotonin levels began to increase for the first time in four years. This upsurge in her dopamine and serotonin reserves was triggered by a sudden influx of oxytocin that now percolated through Curie's brain for the first time since the early years of her relationship with Pierre. Curie had fallen in love.

Curie had known Paul Langevin for some time. At age seventeen he had come to study with Pierre at the Ecole supérieure de Physique et de Chimie and later he taught alongside Curie at the Ecole normale supérieure in Sèvres. He had been great friends with Pierre and was devastated by his death. By July 1910 Curie and Paul had become lovers and would meet regularly in a rented apartment and on those days when they couldn't meet they would exchange passionate letters. But in September a distraught Paul told Curie that his wife Jeanne had stolen her letters to him and was so upset by the affair that it was highly likely that she would send their correspondence to a newspaper in an attempt to publicly disgrace Curie. With these fears in mind Curie and Paul met less often and Paul was more attentive to his wife. But then in November 1911, as Paul and Curie returned to Paris from the Solvay Conference, *Le Journal* published a picture of Curie on its front page under the banner headline : 'A story of love: Madame Curie and Professor Langevin.'

The feature that now filled all the French newspapers was given even more momentum when on 7 November Reuters released the news that Curie had been awarded the Nobel Prize for Chemistry, the first person ever to win a second Nobel Prize. This unique honour seemed to exacerbate Jeanne's fury and she now charged her husband with 'consorting with a concubine in the marital dwelling' – a charge that would be heard in a criminal court. Jeanne's lawyer prepared his case by citing intimate passages from Curie and Paul's letters and the French press had a field day, launching attack after attack against Curie. Eventually the letters were published in the press, igniting public outrage, and Curie was branded 'the foreign woman who was destroying a French home.'

The Nobel Foundation joined the fray and suggested it would be inappropriate if Curie travelled to Stockholm to receive the prize, as it would be an embarrassment for the Swedish king to meet her, given the current circumstances. Curie wrote back: 'The prize was awarded for the discovery of radium and polonium. I believe that there is no connection between my scientific work and my private life.' She then announced that she would attend the ceremony which she did, accompanied by her daughter Irene and sister Bronia, where she was presented to King Gustav who showed no sign of any embarrassment in honouring Curie for the second time in eight years. Somehow, despite her exhausted state, Curie managed to give the customary Nobel lecture with flair and dignity, but soon after her return to Paris she collapsed and remained seriously ill for the next two years. In this diminished physical and psychological state there was no question of continuing her relationship with Paul and it appeared that her scientific work was also at an end. In March 1912 she had a life-saving operation on her damaged kidneys, but she felt sure she was about to die. Aged forty-four it seemed that her prolonged midlife crisis would prove fatal. Weighing only 103 pounds, she had lost all interest in her life and work.

The first signs of any kind of recovery appeared about a year later when Albert Einstein paid her an extended visit. His invigorating company seemed to energise her, to initiate the slow recovery from her depression. He wrote to her: 'I am profoundly grateful to you, as well as to your friends, for having allowed me, during these days, to truly share your existence. I know of nothing more inspiring than seeing beings of your quality living together so perfectly.' The two old friends then decided to share a summer holiday in the beautiful Engadine Valley in Switzerland which proved just the tonic that Curie needed and by the summer of 1914 she was back in her laboratory, continuing new avenues of scientific research.

As soon as the German army invaded the north-east of France, Curie collected her entire stock of radium and travelled with it to Bordeaux where, safely contained in a lead-lined box, she placed it in the vault of one of the city's banks. As the casualties from the First Battle of the Marne limped back into Paris, Curie knew exactly what she should do. Understanding the great diagnostic benefits of X-ray techniques, she designed what she called a radiological car, a large van containing an X-ray machine, photographic darkroom equipment and an electric generator, which could be driven up to the battlefield where army surgeons could use X-rays to help them decide what amputations or other operations were needed. She understood that on the battlefield misdiagnosis was widespread and that many lives could be saved and many amputations avoided by using X-rays.

Inevitably the army authorities were initially opposed to this interfering woman with her strange contraption and they weren't prepared to provide funds to build a fleet of these vehicles, but once again Curie's obstinacy proved formidable. She approached the philanthropic organisation the Union of Women of France who raised sufficient money for the first prototype vehicle and soon, with further donations, dozens of vehicles were built. Curie then learned to drive, change punctured tyres, mastered rudimentary vehicle maintenance and became a dedicated war worker, driving up and down the front, tending to the wounded and recruiting dozens of young women to drive her ever-growing fleet of radiological vehicles, which now become known throughout France as 'little Curies'. In addition, she supervised the construction of 200 radiological clinics that were strategically positioned to tend to the wounded soldiers brought up from the front line.

No other belligerent nation had anything like this humane facility that Curie single-handedly gave the French army. Finally able to acknowledge the outstanding contribution Curie was making to the war effort, the Army High Command opened a school for X-ray technicians in Paris with Curie as director and where 150 army *manipulatrices* – who were mostly women – were trained in the course of its first year. After the great war, Curie became almost as famous for her phenomenal war work, as she was for her scientific achievements. The people and government of France seemed to forget the scandal of her affair with Paul Langevin and awarded her a pension for life.

But as if that weren't enough, a final consequence of Curie's midlife crisis was her momentous contribution to the women's movement, which was demanding changes at all levels of society. As a woman who had broken through so many bastions of male dominance, Curie became a global role model for an emerging feminism and devoted her life to its causes. Indeed, the manner in which she had led the way, forging an array of new opportunities for women in France, was celebrated in the spring of 1921, along with her war-work and scientific achievements, in a gala evening at the Paris Opera, hosted by the French president. The cast of eminent guests included the great actress Sarah Bernhardt, who read a tribute to Curie entitled 'Ode to Madame Curie', hailing her as the 'sister of Prometheus'.

Curie's crusade for women's rights then continued in a nationwide tour of America where a few months earlier women had been granted universal suffrage. She visited and spoke at women's colleges and universities throughout the country and the visit reached its climax when 3,500 women gathered at Carnegie Hall in New York to celebrate Curie's contribution to international feminism. This was followed by a grand reception at the White House where President Harding presented her with a gram of radium as a token of America's women's gratitude for her on their behalf.

On her return to Paris, and indeed for the rest of her life, much of her time was devoted to the establishment and development of the Institut du radium at the Sorbonne, which employed a significant number of women and which soon ranked as one of the world's leading scientific institutions.

Once she had recovered from the agonies of her midlife crisis, Curie's final twenty years were filled with remarkable achievements. One is left to wonder whether her further scientific work, her founding of some of the world's most prestigious scientific institutions, her pioneering humane work on the Western Front during the war and, perhaps most important of all, her contribution to the global women's movement would have been achieved if Pierre had lived. The crisis of her middle years fashioned a woman of such forceful personality and moral distinction that it enabled her to evolve beyond the brilliance of her scientific work to become an iconic humanitarian and radical, groundbreaking feminist.

– Crisis, Creativity and the Work of Heinz Kohut –

As we saw in chapter four, Beethoven's 'Heiligenstadt Testament' describes a common feature of the midlife crisis: when a growing realisation that an illness, an addiction, a failing relationship, a professional collapse, an inner mood or tension has slowly increased from an initial apprehension to the full certainty of its disastrous impact on the individual's life. But Beethoven's defiance against his cruel fate, as his deafness became complete, produced a sequence of final masterpieces which have never been surpassed. His late piano sonatas, his ninth symphony, his Missa Solemnis and perhaps most of all, his late string quartets represent a body of music which touches a level of profundity that has seldom been matched before or since. As we also saw with Roosevelt, illness is frequently a life-changing, midlife experience, but the midlife crisis can also be triggered by exhaustion caused by chronic overwork resulting in what we now call 'burnout'. This physical and mental collapse was experienced by Michelangelo and Tolstoy, which is hardly surprising given the monumentality of their artistic aspiration that resulted in the ceiling of the Sistine Chapel and *War and Peace*.

This prodigious level of ambition was forged in Michelangelo and Tolstoy's psyches by a state of mind that the leading psychoanalytic theorist Heinz Kohut has called 'grandiosity'. Kohut regards the infant's view of the world as a very localised, individualised environment where he or she is omnipotent and this sense of omnipotence will create what Kohut calls the 'grandiose self', characterised by 'an all-embracing narcissism by concentrating perfection and power upon the self.' This powerful sense of self-worth is an important, healthy starting point in life, but the parents must help the child recognise and accept its limitations and the parameters that surround the child. This reality check depends upon optimal parenting, but if this reality regulation isn't achieved and the necessary moderation of the grandiosity doesn't take place, this feature of the developing psyche will be repressed and can easily become troublesome in adulthood, ranging from cases of delusional narcissistic grandiosity to more common features of what Jung called 'inflation'. However, Kohut writes that very gifted individuals may be driven to their greatest achievements by the demand of 'a poorly modified grandiose self'.

It is therefore a salient aspect of their psychobiography that Tolstoy's mother died when he was two and that Michelangelo lost his mother aged four and one is left to wonder whether both these individuals produced these monumental artistic creations because they lost this critical maternal reality regulation at a very early age. In each case this emotional catastrophe resulted in an adult depressive nature which they attempted to relieve by using this psychological strategy of grandiosity and in both their lives this tension between feelings of low self-esteem and hopelessness and the vaulting ambition of grandiosity was an essential feature of their respective midlife ordeal.

In October 1508 Michelangelo began work on the ceiling of the Sistine Chapel. Fifteen months later he felt crippled by the immensity of the task he had set himself and in a state of morose despondency he wrote to his father that he was suffering from 'the greatest physical fatigue', lamenting the fact that he had not painted in fresco since the days of his apprenticeship and complaining: 'It's not my trade.' In a letter to his friend Giovanni da Pistoia he wrote: 'This miserable job has given me a goitre like the cats in Lombardy . . . Please take up my cause, defend both my dead painting and my reputation. I'm in a poor state and I'm not a painter.' Hounded and intimidated by his fearsome patron Pope Julius II, Michelangelo felt that accepting the Sistine Chapel ceiling commission had been an act of arrogant hubris. The whole enterprise now overwhelmed him and he felt sure that he was destined to suffer a final humiliation which would end his career.

Throughout much of 1510 he remained in this state of disconsolate depression. There seemed no one to support him, no one to encourage him as he toiled away on his scaffold. The daily painful contortions involved in his work, with his arms and hands continually above his head, caused him both eye strain and back pain, and added layer upon layer of exhaustion onto his mind and body. To make matters worse Julius II was now away leading the Italian-Swiss army which he had mobilised to face Louis XII's French troops that occupied Genoa, Milan, Verona and Ferrara. Michelangelo agonised over the possibility that if Louis defeated Julius, French soldiers would then pillage and loot Rome and destroy the work he had already accomplished. To add insult to injury Julius hadn't paid him for months. Finally, in September, Michelangelo travelled to Bologna in a state of despair to confront the Pope, but with no success. More miserable than ever Michelangelo returned to Rome refusing to continue his task until he had been paid. For five months he did nothing, incapacitated by his depression, but within this enforced inactivity an unseen creative gestation was developing.

After months of inertia, he started working on tiny drawings, done in red chalk, preliminary sketches on small pieces of paper measuring seven by ten inches. The emergence of these seemed to soothe his anxieties and quicken the pulse of his creativity and as the studies developed, he used them to complete a large cartoon entitled 'The Creation of Adam'. By September 1511 he had been paid and the necessary scaffolding was erected and by early October, he was working at a furious pace on his new Sistine fresco. Miraculously by late November 'The Creation of Adam' was complete. In the execution of the central image of the Sistine commission he had succeeded beyond even his own expectations, providing us with one of the greatest masterpieces in the entire canon of Western art. It appears that this prolonged absence from the Sistine Chapel allowed Michelangelo the full experience of his midlife crisis and his transformatory recovery would result in his crowning achievement and the sublime works of his maturity.

Tolstoy finally finished *War and Peace* in the spring of 1867, after toiling away at it for five years. His mood, however, was far from elated, despite the immediate success of his epic novel. He felt listless, depleted both physically and mentally, and although he was only forty-one he become excessively preoccupied by his own mortality. To avert his attention away from these circling thoughts of death he decided to use money earned from *War and Peace* to purchase an estate in distant Pensa. In September 1867 he travelled by train to Nizhny Novgorod where he hired a coach to complete the journey. Tired and despondent, he arrived in the small town of Arzamas where he spent the night, during which he awoke in a state of agitated panic. Three days later he wrote to his wife: 'I was overcome by despair, fear and terror, the like of which I have never experienced before. I've never felt such an agonising feeling before and may God preserve anyone else from experiencing anything like this.'

What Tolstoy described is one of the most common features of the midlife crisis – an overwhelming fear of death often felt by high-achievers who during their middle years, when their ambitions have been fulfilled, feel that all their achievements have been pointless and of no value because in a handful of years they will have to face death, a dreadful reality that will render their success as nothing more than meaningless vanity.

It was this challenge that Tolstoy was now facing, just at the very moment when *War and Peace* was receiving national and international acclaim. He returned empty-handed from his journey to Pensa gripped by this mixture of terror and depression that was to remain with him for months to come. The extent of his torment was described in a short story entitled 'Memoirs of a Madman', where he wrote: 'I am always with myself, I am my own torturer. Here I am the whole me. It is myself that I am tired of and whom I find intolerable and a torture. I want to fall asleep and forget myself but I can't. I can't get away from myself.' This self-loathing was the reverse side of Tolstoy's grandiosity which had provided the psychological drive and superhuman energy that results in such accomplishments as *War and Peace* or the ceiling of the Sistine Chapel.

It was Tolstoy's midlife ordeal that resulted in his next book *Anna Karenina*, regarded by many as a more perfect novel than *War and Peace*. Yet more specifically it was this experience that began Tolstoy's transformation from a writer of fiction into an ethical and spiritual leader whose thoughts and writings were to have a profound impact in Russia and around the world, deeply influencing individuals as diverse as Gandhi and Wittgenstein. In fact throughout the Great War Wittgenstein carried with him only one book, Tolstoy's *The Gospel in Brief.* He became known, while in the trenches, 'as the man with the Gospels' and he wrote to his friend Ludwig von Ficker from the Eastern Front: 'If you are not acquainted with [Tolstoy's book], then you cannot imagine what effect it can have on a person. At this time this book virtually kept me alive.'

A common cause of the midlife rite of passage is when an individual offends the values and ethics of the collective that surround them in such a manner that causes them hardship and suffering, even if a sense of liberation may be achieved. In addition, the midlife ordeal, as we have seen, almost always provides the individual concerned with, as it were, two lives – a life before the crisis and a quite different life after the crisis. This happened most dramatically to Mary Ann Evans.

In May 1849 she lost the person whom she loved most. 'My father was the one strong deep love I have ever known. What shall I be without him,' she wrote to her friend Charles Bray. As she sat with her beloved father as he lay dying, she was reading a book entitled *The Soul*, by Francis Newman: 'There are two families of children on this earth: the once born and the twice born.' Slightly amending Newman's words I think it would be true to say that people who avoid the midlife crisis have one life; people who go through the midlife crisis have two. But often there is a gap or as we have seen, a period of liminality between one life ending and another life beginning: the period of critical transformation that prolongs the midlife transition from its sudden traumatic opening event, through the period of prolonged liminality to the point when the new second life begins. Reading *The Soul* Mary Ann Evans knew that her first life was coming to an end. Seven years later she wrote: 'September 1856 made a new era in my life, for it was then that I began to write fiction.' During these seven years of liminality she faced much adversity, changed from being Mary Ann Evans to George Eliot and became, like Marie Curie, a pariah who was ostracised by her friends, family and the society and culture of her country.

Mary Ann Evans suffered all her life from feelings of inferiority because throughout her childhood her mother constantly referred to her plain, unprepossessing physical appearance. Fourteen months after Mary Ann's arrival her mother gave birth to twin boys, who both died within a few days of their birth, casting her mother into a deep depression from which she never recovered. From then on, her mother seemed to resent her youngest child, always preferring Mary Ann's other siblings.

In my practice I have noticed that when a child is unable to attract sufficient attention from a distracted, unsympathetic parent, they can develop a considerable natural charm which is then used to attract surrogate parental figures who can fill the gap left by the disinterested biological parent. Mary Ann did exactly this. In the years following the loss of her father three older men became enchanted by her, the first of these being the famous painter François d'Albert-Durade; the second John Chapman, who made her the editor and weekly contributor of his *Westminster Review*, one of the leading literary periodicals of the day; the third, writer Herbert Spencer, with whom she fell deeply in love. In all three cases, despite her wit, intelligence and engaging company, none of these men were sexually attracted to Mary Ann and the inevitable emotional disappointment cast her into despair and depression until she met George Lewes in Jeff's bookshop in the Burlington Arcade on 6 October 1851. The meeting was to change not only her life, but also her name and profession. At the same time, however, this relationship would scandalise her family, friends and the society in which she lived because even though George Lewes was separated from his wife Agnes they remained married. She was therefore confronted by the greatest challenge of her life as her true authentic needs found themselves in defiant opposition to all the collective values she was surrounded by. From early 1853 until the end of Lewes's life they were virtually inseparable, although they never married. Agnes lived with Lewes's business partner, Thornton Hunt, with whom she had further children and Lewes agreed that these children could be registered as his own to avoid the stigma of illegitimacy. By doing this, he was legally condoning his wife's adultery and consequently he was, by law, barred from ever applying for a divorce, which in turn meant he could never marry Mary Ann unless Agnes died.

As 1853 progressed, news of Mary Ann's relationship with Lewes began to circulate and the first two people to question her judgement were her two greatest friends Sara Hennell and Cara Bray. In addition, her family also took exception to what they saw as an act of moral degeneracy. Particularly offended was her brother Isaac, to whom she had been close and who, after much recrimination, refused ever to see her again and communicated with her only through lawyers.

Mary Ann was overjoyed that at last she had found someone she loved, who entirely reciprocated her feelings both physically and emotionally, yet she suffered waves of depression as she tried to adjust to the withdrawal of the approval and affection that she had enjoyed and which her fragile nature still needed so much. She was now plagued by a spate of illness and physical ailments, suffering from coughs, a constant sore throat and rheumatic pain, which required weeks of bedrest. She was afflicted by severe digestive troubles and persistent toothache. Providence had confronted her with an agonising predicament for if she was to stay with the man she loved she would never be able to marry him, but living in sin with George Lewes would bring down on her the full denunciation of a vindictive Victorian society.

Such irreconcilable circumstances have been named by the anthropologist and psychologist Gregory Bateson as 'the double bind', a predicament I continually see in my client work. As described by Bateson, a double bind is a dilemma in which the individual is faced by two conflicting situations, where one negates the other. This creates a situation in which a successful response to one need results in a failed response to the other need. These double-bind circumstances are often a salient feature of the midlife crisis, when a client attempts the developmental necessity of moving from their old, depleted life into a vibrant new life. Similarly the double bind usually pitches the personal developmental needs against the needs of others whom the client is responsible for and becomes the common ignition point of the midlife crisis. This was what Mary Ann was now faced with: whether to comply with the moral, ethical expectations of friends and family whom she loved and end her romantic attachment to George Lewes or to ignore these constraints and commit herself fully to this unconventional and morally unacceptable relationship. She chose the latter of these two stark alternatives. For Mary Ann the situation in London had become unbearable and she and

Lewes escaped to Germany. Even though she hadn't announced her plans to those closest to her, the news of their 'elopement' circulated around her friends and acquaintances in London and Edinburgh almost immediately. Mary Ann had crossed her Rubicon. She was on her way to becoming George Eliot, despite being widely described as 'George Lewes's concubine'.

After nine months travelling in Germany the couple returned to England to face the full wrath of censorious Victorian society where they would remain outcasts and pariahs for the next decade until finally the full stature of George Eliot, both as a woman and as a writer, emerged from this long crisis. As *Adam Bede*, *The Mill on the Floss*, *Silas Marner* and finally *Middlemarch* took their place amongst the great novels of the day Victorian society, led by the Queen herself, forgave George Eliot her earlier indiscretions.

Heinz Kohut suggests that many of us, in our early adulthood, design a kind of blueprint of goals and ambitions which remain with us, mostly unrealised, until middle age when some event turns our attention back to these hopes and we find that life's course has fallen considerably short of our original aspirations. In *The Restoration of the Self*, he states that in middle age, when we're moving inexorably towards our ultimate decline, 'we ask ourselves whether we have been true to our innermost design.' We still have the time and energy to put things right before we face the precipice of old age, but these shortcomings in our hoped-for aspirations cause us perturbation and we may feel the need, in our middle years, to make some drastic adjustments to our life's course.

I had a client named Ben who had been practising as a counsellor for many years. He came to see me feeling something was awry: that something needed urgent attention which he couldn't quite name but which made him feel listless and unsettled. Six months into therapy he relayed a dream to me in which his house was on fire. He felt compelled to enter the burning building to rescue three prized possessions – his wedding ring, his binoculars and a file full of poems that he had written when he was a young man. His wedding ring represented his marriage, upon which he placed great value. The binoculars symbolised his love of nature, particularly his enthusiasm for birdwatching and his commitment to eco-activism. Both these aspects of his life we had discussed and examined thoroughly, yet the file of poems he had never mentioned, indeed he had almost forgotten about their existence. He then told me that when he was a young man he had ambitions to become a poet and had dozens of poems lying forgotten at home. I asked him to bring some of these to our sessions: the dream told us that these poems needed to be looked at again. When I saw what he had written I was impressed by their philosophical content and his obvious lyrical gift. Together we began to discuss and examine how these poems were a kind of Kohutian blueprint laid down during his early adulthood that had remained out of view for thirty years but still had a powerful presence in his psyche. As we explored these early ambitions, and why they had been repressed into his unconscious, he started to write again. A cascade of poems streamed from his pen, many of which seemed to me to be of high quality.

Ben's parents had given him very little encouragement during his childhood and adolescence. Now suddenly his poetic gift had announced itself by means of this dream just at a time when I, as a paternal surrogate, could give him all the encouragement he needed to nourish his undoubted poetic gift that now took its rightful position at the centre of his life.

We all have the capacity within us to express our innate creativity. This creative urge, which often doesn't receive the encouragement needed to develop its fragile nature, can – if rediscovered during midlife – provide us with great sustenance and fulfilment during our mature years.

– Synchronicity and the Gravitational Field of Authentic Need –

One of the great intellectual adventures of Carl Jung's life was his long relationship with the Nobel Prize-winning physicist Wolfgang Pauli who progressed from being Jung's patient, to friend and then to collaborator and co-author. This highly productive partnership produced a body of groundbreaking work that fused Jung's depth psychology with Pauli's quantum physics. This resulted in their joint creation – synchronicity – a concept that plays a significant role in midlife psychology.

Wolfgang Pauli was born in Vienna in 1900. His father was Professor of Chemistry at Vienna University and his godfather was the eminent physicist Ernst Mach. At school Pauli was a brilliant student who relieved his boredom during lessons by studying Einstein's latest papers on relativity. In 1918 he went to Munich University where he submitted a paper on general relativity which was praised by Einstein and established him – despite his youth – as a leading figure in European scientific circles. While in Munich he befriended both Werner Heisenberg and Niels Bohr and subsequently moved to Copenhagen where together these three great physicists laid down the foundations of quantum mechanics. By 1928 Pauli moved to Zürich University where he became Professor of Physics.

Achieving so much at such a young age appeared to have a destabilising effect on Pauli. In addition, his family harboured a shameful secret that most likely affected him. His father had had numerous extramarital affairs which had resulted in his mother's suicide and by the time Pauli was settled in Zürich he was – like his father – leading a secret double life. By day he behaved like a staid professor, but by night he roamed Zürich's red-light district, with its dingy bars and seedy brothels. Pauli's late nights would often end with him being beaten up in some brawl and he would return to his lab the next morning bruised and battered, yet strangely able to continue his demanding scientific work. This Dr Jekyll-and-Mr Hyde existence suggests that Pauli had an unconscious understanding that regular contact with his shadow would intensify his creative output.

On one of his nightly frolics he met a chorus girl called Käthe Deppner and within four weeks he impulsively proposed marriage. After the wedding their relationship immediately fell apart and Deppner walked out on Pauli, which led to yet another year of excessive drinking, cruising Zürich's red-light district and further bar-room brawls. As this early midlife crisis escalated, his work at the university ground to a halt and finally his concerned colleagues persuaded him to consult Dr Jung who lived just down the lake from where Pauli lived.

In preparation for his sessions with Jung, Pauli read almost all of Jung's published works and Jung was amazed how this physicist had grasped the complexity of his theories. After several years of intense therapy Pauli appeared to have recovered from his midlife crisis, but both he and Jung wanted their fruitful relationship to continue and so they developed a partnership that produced a number of significant papers and books which included most notably *The Interpretation of Nature and the Psyche* and *Synchronicity: An Acausal Connecting Principle.*

Put simply, in their work on synchronicity Jung and Pauli suggest that the hidden atomic structure of a physical entity has similarities to the unseen structure of the unconscious, with the nucleus of the atom similar to the Self. Both men were convinced that the concept of synchronicity occurred when a piece of unconscious material moved from the unconscious to the conscious which then causes a transformation in the psyche. As this happens the Self emits a plume of energy, like a pulse of electricity, which attracts a corresponding outer physical event. A common example of this phenomenon in real terms is when a client becomes conscious of a defence that he or she was not previously aware of which has prevented him or her from sustaining a committed adult relationship. Once this defence is explored and the early trauma that caused it is understood, the client then often meets someone who turns out to be their eventual life partner. When this happens the coincidence is so dramatic that it reveals the connection between the psychological and the physical event.

Over many years I have spoken to numerous friends and colleagues about the strange coincidences and serendipitous experiences that we all encounter from day to day. From this anecdotal evidence there seems to be a consensus that a number of these experiences appear to go beyond mere coincidence. Their pattern, shape and frequency feel like some phenomena that breaks the bounds of random chance. Yet the rational pragmatism of our culture stigmatises this kind of evaluation as a flaky, hair-brained attempt to turn a mere coincidence into some mystical proof of telepathy or, even worse, into some kind of specious necromancy.

In his seminal work *The Master and His Emissary* Iain McGilchrist writes of the polarity which stands at the core of our cerebral functioning – the sharply contrasting division between our left- and right-brain hemispheres. What McGilchrist is alerting us to is the dominance of our left-brain rational pragmatism and how this results in our culture's preference for utility, logical consistency, technological dominance and an understanding of reality based purely upon empirical evidence. McGilchrist urges us to pay much more attention to the muted presence of the right brain, with its predisposition for the intuitive, the spiritual, the archetypal, the primal, the feral, the emotional and the unconscious.

McGilchrist contends that the 'master' is the right brain, with all its instinctive qualities, while viewing the left brain and its properties as the right brain's 'emissary', suggesting that the right brain's innate wisdom – because of its 'tentativeness' – needs the help of the streetwise left brain, which he describes as a kind of high achieving bureaucrat. Both aspects are essential, but more balance, both individually and culturally, is needed.

It is often during the questioning and introversion experienced during midlife liminality that we notice a number of coincidental events that Jung and Pauli have named 'synchronistic'. We also begin to listen more to our right brain's instinctive responses and become less driven by our over-active left-brain rationalism. But what mechanism in the psyche empowers us to attempt to dislodge and shed our complexes, paying more heed to synchronistic events and to right-brain intuitive reflections? Working with clients, time and again I see the emergence of 'authentic needs' which stand in sharp contrast to our 'narcissistic needs', those that wish continually to shore up our trauma defences and to comply to the collectives that surround us. This is the principal experience of the midlife crisis. At such times the psychotherapist becomes a kind of psychic geologist excavating the emotional geology of the client for both traumatic wounds and signs of developing individuality. This is obviously a painful, exhausting experience, but it will be aided by a facilitating capacity in our psyches that I have come to call the 'gravitational field of authentic need'.

In our middle years our inner authentic self, activated by surrounding experiences of midlife turmoil and change, builds up enough energy or 'critical mass' to break through ingrained patterns of defence and collective constraints. But to achieve this breakout it has to attract assets and resources that can support and energise a reconfiguration of both our inner and outer lives and it does this by creating a gravitational field which attracts the experience or person that can provide exactly the right encouragement to support the emerging authentic self. I have witnessed both in my work with clients and in my own life synchronistic events which draw exactly the right experience or person into our lives to support us at this critical developmental moment.

One frequent example of how this gravitational field of authentic need functions is the manner in which clients so often find their way to exactly the right therapist at exactly the right time. This was certainly the case in Jung and Pauli's relationship. Jung's primary aspiration throughout his life was to anchor his psychological theories within the context of an accepted and disciplined science. He regarded his thought and writing on synchronicity as having the greatest scientific value and yet he knew that the scientific community would not tolerate his theories of non-empirical speculation. His collaboration with Pauli was a prime example of Jung's deep authentic need attracting and consolidating his relationship with this world-famous physicist. For Pauli his long relationship with one of the two founders of depth psychology was literally a lifesaver. His capacity for self-destructive behaviour would, I suspect, have been terminal had the gravitational field of authentic need not drawn him towards Jung, perhaps one of the very few individuals who had the intellectual power and intuitive capacity to engage Pauli in the process of his recovery from his deep and life-threatening psychological malaise.

I have had a number of similar experiences in my clinical work. One particular client arrived for his first session in a very unsettled and anxious state just before he was about to retire from a long career as a leading acoustician. As we started our work together it became apparent that he felt that he had underachieved in his job. During his account of his working life he described all the concert halls he'd worked on and as a concert promoter myself, during a long career which has run parallel to my psychotherapeutic work, I was familiar with them all. From his depressed state of mind, anchored in a posture of low self-esteem and aggravated by a fierce inner critic, he was completely unable to evaluate objectively the success of his career. However, I knew only too well the quality of the fine acoustics of the halls where he had been engaged as the acoustician. We were therefore able to consider at length the disparity between his own flawed judgement upon his life's work and my knowledge of his masterful achievements.

With this disparity lying so blatantly before us, we then explored the constricting complex that forbade him from viewing his career favourably and we were able to loosen its constricting impact. After several years – as his acerbic inner critic became less vocal – his low self-esteem slowly evaporated, which allowed us in the last stages of our work together to celebrate the very considerable success of his career. I have often pondered on how this master acoustician found his way to me, almost certainly the only psychotherapist in the world to have a deep knowledge of the acoustics of British concert halls. For me this remarkable coincidence is an example of the gravitational field of authentic need.

I have also been aided by such synchronistic experiences. For many years I had wanted to visit Jung's tower at Bollingen, however I had no idea where the tower was located, nor indeed how I might gain access to the private land that surrounded it. Despite my ignorance, I decided to fly to Zürich with one of my daughters in April 2016 and see what happened. Then three weeks before our departure a friend told me that someone was giving a lecture about Jung's tower. I attended and when it was over I introduced myself to

Martin, the man who gave the lecture, and told him that I was setting off to visit Bollingen and did he have any advice to give me. After a couple of minutes' conversation he suddenly announced he would clear his diary and come to Switzerland with us to act as our guide. This resulted in a life-changing trip which included a tour around the Burghölzli psychiatric hospital where Jung worked, a visit to his house in Küsnacht, to the C. G. Jung Institute, to his grave and finally on a beautiful spring day we spent four memorable hours at the tower itself. Somehow the strength of my authentic need drew me to my first encounter with Martin and our subsequent journey to Bollingen, an experience that has had a deep impact upon my inner development.

McGilchrist's emphasis on the importance of more right-brain intuitive development strengthens Jung's concept of individuation as a process central to human evolution, invigorating the need for a much more effective partnership between relational ethics and mechanistic technology. The immense increase in individual desire for psychological development is surely a reflection of our species' collective aspiration to find global adaptions which will help us overcome the environmental challenges we now face. At present, unless the altruistic, ethical side to our nature restrains our drive for technological development, the nationalistic, tribal cultures which were so destructive in the twentieth century will prevail, resulting in ecological disaster. It is therefore not too much to say that individuation is critical to the current evolutionary challenge that our species now faces. To provide a sufficient number of individuals needed to create a critical mass of right-brain ethical wisdom, the process of midlife development has to be widely present, as does its essential catalyst: the midlife crisis.

– Conclusion –

The work of Freud, Jung, Klein, Winnicott, Kohut and Grof place an onerous pressure upon parents, and particularly mothers, to provide their children with a full measure of love, empathy and understanding. And yet I've noticed time and again that those individuals, whose parents have given them this high level of attention and love, often live in such a reliable, benign world that a frequent consequence of their agreeable emotional environment is that they lack ambition, drive and a certain creative edge.

Perhaps human evolution requires a degree of psychological dissonance to provide sufficient individual energy, motivation and drive to propel society forward. Indeed those individuals who have contributed most to our collective progress tend to have experienced a level of psychological dysfunctionality that furnishes them with enterprise, determination and tenacity. It is as if the traumas of birth and early childhood, the mismatch between the amygdala and hippocampus during infancy and the deficiencies in optimum parenting can sculpt a kind of compensatory impetus and thrust in the individual, which collectively advances our culture's progress to increasing levels of achievement.

And yet humanity's chaotic and often brutal history shows us that all this febrile energy carries with it an ominous shadow which in recent decades has become so contorted and egregious that it may result in some extinction event. It is this threat which the potential wisdom, empathy and altruism of the second half of life can hopefully control. As we have seen, through the challenges and ordeals of midlife, a number of exceptional individuals have emerged who have reached a level of emotional and ethical maturity, which has given them the qualities needed to steer our culture through a series of historic calamities. This is why I believe there is an evolutionary purpose to this period of potential transformation in our middle years that we've come to call the midlife crisis.

I would like to finish with the suggestion that we are now in the middle of humanity's midlife crisis, which will perhaps run for 200 years, as we struggle to survive a series of potential extinction events. The process began with the horrors of the first half of the twentieth century which were followed by our escape from nuclear annihilation by the narrowest of margins and now we are being tested by the increasing devastation of global warming. It seems to me that the events of the twentieth century have made us question the eternal verities that underpinned the previous 2,000 years and the authority and reliable certainties of religion, the nation state and the family are now seen to be full of flaws and inadequacies. As we continue to struggle with these unprecedented challenges, our global midlife crisis can only be overcome if humanity discovers some new collective way of being that safely guides us into some unimagined future.

– Acknowledgements –

This book could not have been written without long conversations with many friends and the love and support of my wife and three daughters, Julia, Anna, Lucy and Olivia. The garnering of the manuscript could not have been completed without the endless patience of Fiona Todd, who made up for my digital deficiencies, and the masterful and indispensable encouragement of my editor Rosalind Porter. Over a decade or more invigorating talks with Judith Griffin, Martin Gledhill and Rob Weston have helped me gather my thoughts on the subjects discussed in these pages. I have, however, learned most from my clients (whose names have been changed) to whom I have often wanted to say: 'My turn to pay this week.'

Other titles from Notting Hill Editions*

Brainspotting: Adventures in Neurology
A. J. Lees

As a trainee hospital doctor, A. J. Lees was enthralled by his mentors: esteemed neurologists who in their work combined the precision of mathematicians with the solemnity of undertakers. Today their clinical methods honed at the bedside are in danger of extinction, replaced by a slavish adherence to algorithms, protocols, process and a worship of machines. In this series of brilliant autobiographical essays, Lees takes us on a grand tour of his neurological career giving the reader insight into the art of listening, observation and imagination that the best neurologists still rely on to heal minds and fix brains.

On Cats: An Anthology
Introduced by Margaret Atwood

Introduced by Margaret Atwood, who was a 'cat-deprived young child', the writers in these pages reflect on the curious feline qualities that inspire devotion in their owners, even when it seems one-sided. Includes contributions from Doris Lessing, Edward Gorey, Mary Gaitskill, Ernest Hemingway, Caitlin Moran, Nikola Tesla, Muriel Spark, John Keats, Lynne Truss, Guy du Maupassant, Rebecca West, Hilaire Belloc and more.

The Wrong Turning: Encounters with Ghosts
Introduced and Edited by Stephen Johnson

Why do people love ghost stories, even when they don't believe (or say they don't believe) in ghosts? With contributions from M. R. James, Alexander Pushkin, Charlotte Perkins Gilman, Tove Jansson and more, this uniquely curated anthology brings together some of the most chilling stories from around the world.

You and Me: The Neuroscience of Identity
Susan Greenfield

What is it that makes you distinct from me? Identity is a term much used but hard to define. For that very reason, it has long been a topic of fascination for philosophers but has been regarded with aversion by neuro-scientists – until now. Susan Greenfield takes us on a journey in search of a biological interpretation of this most elusive of concepts, guiding us through the social and psychiatric perspectives and ultimately into the heart of the physical brain. As the brain adapts exquisitely to environment, do the cultural challenges of the 21st century mean that we are facing unprecedented changes to identity itself?

The Mystery of Being Human: God, Freedom and the NHS
Raymond Tallis.

For forty years, Raymond Tallis was an NHS consultant. In this brilliant collection, Tallis brings his signature intelligence and razor wit to the questions that define us as human: Do we have free will? Can humanity flourish without religion? Will science explain everything? And can the NHS – an institution that relies on compassion over profit – survive?

How Shostakovich Changed my Mind
Stephen Johnson
Winner of the 2021 Rubery Book Award

In this powerfully honest and brilliant book, BBC music broadcaster Stephen Johnson explores the impact of Shostakovich's music during Stalin's reign of terror and writes – at the same time – of the extraordinary healing effect of music on the mind. As someone who has lived with bipolar disorder for most of his life, Johnson looks at neurological, psychotherapeutic and philosophical findings, and reflects on his own experience of how Shostakovich's music helped him survive the trials and assaults of mental illness.

Questions of Travel: William Morris in Iceland
Lavinia Greenlaw

The great Victorian William Morris was fascinated by Iceland, which inspired him to write one of the masterpieces of travel literature. In this fascinating book, which is part memoir, part prose poem, part criticism and part travelogue, celebrated poet Lavinia Greenlaw follows in his footsteps, combining excerpts from his Icelandic writings with her own eye-witness response to the country.

Mourning Diary
Roland Barthes

Roland Gérard Barthes (1915–1980) was a guru among literary theorists. Raised by his mother, Henriette was the most important person in his life, and yet Barthes's devotion to her was unknown to even the closest of his friends. Written as a series of notes on index cards, as was his habit, *Mourning Diary* shows us how Barthes began reflecting on a new solitude only days after his mother's death by recording the impact of bereavement as he struggled to live without her. The result is 330 cards that are at once intensely personal and entirely universal.

Wandering Jew: The Search for Joseph Roth
Dennis Marks

Joseph Roth, whose many novels included *The Radetzky March*, was one of the most seductive, disturbing and enigmatic writers of the twentieth century. Born in the Habsburg Empire in what is now Ukraine, and dying in Paris in 1939, he was a perpetual displaced person, a traveller, a prophet, a compulsive liar and a man who covered his tracks. In this revealing 'psycho-geography', Dennis Marks makes a journey through the eastern borderlands of Europe to uncover the truth about Roth's lost world.

Mentored by a Madman: The William Burroughs Experiment
A. J. Lees

A fascinating account by one of the world's leading neurologists of the profound influence of *Naked Lunch* author, and drug addict, William Burroughs on his medical career, and on the discovery of a ground-breaking treatment for Parkinson's Disease.

Alchemy: Writers on Truth, Lies and Fiction
Introduced by Iain Sinclair

Reality versus fiction is at the heart of the current literary debate. We live in a world of docu-drama, the 'real life' story. Works of art, novels and films are frequently bolstered by reference to the autobiography of the creator, or to underlying 'fact.' Where does that leave the imagination? And who gets to define the parameters of 'reality' and 'fiction' anyway? In this riveting collection, introduced by one of Britain's most celebrated writers, Joanna Kavenna, Gabriel Josipovici, Benjamin Markovits, Partou Zia and Anakana Schofield debate.